Limits of Indeterminacy in Measure of T-means of Subseries of a Trigonometric Series

1981 subscription consists of 4 issues, published quarterly by American Mathematical Society. Subscription price of 1981 is list $210.00, member $105.00. Previously translated Russian Numbers are sold as separate books. Write to the Society for prices.

Subscription and orders should be addressed to American Mathematical Society, P. O. Box 1571, Annex Station, Providence, R. I. 02901. *All orders must be accompanied by payment.* Other correspondence should be addressed to P. O. Box 6248, Providence, R. I. 02940.

Proceedings of the Steklov Institute of Mathematics (ISSN 0081-5438) is published quarterly by the American Mathematical Society at 201 Charles Street, Providence, Rhode Island 02904. Second Class postage paid at Providence, Rhode Island 02940. Postmaster: Send address changes to Proceedings of the Steklov Institute of Mathematics, American Mathematical Society, P. O. Box 6248, Providence, RI 02940.

September 1981

Proceedings
of the
STEKLOV INSTITUTE
OF MATHEMATICS

1981, ISSUE 3

Limits of Indeterminacy in Measure of T-means of Subseries of a Trigonometric Series

by

D. E. Men´šov

Translation of
ТРУДЫ
ордена Ленина
МАТЕМАТИЧЕСКОГО ИНСТИТУТА
имени В. А. СТЕКЛОВА

Том 149 (1978)

АКАДЕМИЯ НАУК

СОЮЗА СОВЕТСКИХ СОЦИАЛИСТИЧЕСКИХ РЕСПУБЛИК

ТРУДЫ

ордена Ленина

МАТЕМАТИЧЕСКОГО ИНСТИТУТА

имени В. А. СТЕКЛОВА

CXLIX

Д. Е. МЕНЬШОВ

ПРЕДЕЛЫ НЕОПРЕДЕЛЕННОСТИ ПО МЕРЕ T-СРЕДНИХ ПОДРЯДОВ ТРИГОНОМЕТРИЧЕСКОГО РЯДА

Ответственный редактор (Editor-in-chief)
академик С. М. НИКОЛЬСКИЙ (S. M. Nikol'skiĭ)

Заместитель ответственного редактора (Assistant to the editor-in-chief)
доктор физ.-матем. наук Е. А. ВОЛКОВ (E. A. Volkov)

издательство "наука"
Москва 1978

Translated by R. P. BOAS

ABSTRACT. This monograph is devoted to the problem of the representation of functions by trigonometric series. In the very general situation when the functions under consideration may even take infinite values on sets of positive measure, the author establishes the existence, for a given regular summation method, of a universal trigonometric series, with coefficients tending to zero, which represents, in the sense of summability in measure of appropriate subseries, all measurable functions.

The monograph will be of interest to specialists in the theory of functions and related parts of analysis, and to graduate and postgraduate students in mathematics.

1980 *Mathematics Subject Classification.* Primary 42A24.

Library of Congress Cataloging in Publication Data

Men'shov, D. E. (Dmitriĭ Evgen'evich), 1892–

Limits of indeterminacy in measure of T-means of subseries of a trigonometric series.

(Proceedings of the Steklov Institute of Mathematics, ISSN 0081-5438; 1981, issue 3)

Translation of: Predely neopredelennosti po mere T-srednikh podri͡adov trigonometricheskogo ri͡ada.

Number in Russian series statement: no. 149 (1978)

Bibliography: p.

1. Fourier series. I. Title. II. Series.

QA1.A413 no. 149 [QA404] 510s 81-14992

ISBN 0-8218-3043-0 [515'.2433] AACR2

Translation authorized by the All-Union Agency for Author's Rights, Moscow

§1. **Introduction.** We consider an arbitrary infinite matrix

$$\|c_{mk}\| \qquad (m, k = 0, 1, 2, \dots) \tag{1.1}$$

and any series of numbers

$$\sum_{j=0}^{\infty} u_j \tag{1.2}$$

with partial sums s_k $(k = 0, 1, 2, \dots)$. We say that the series (1.2) is *summable to the value t by the method T* defined by (1.1) if, first, the series

$$\sum_{k=0}^{\infty} c_{mk} s_k \tag{1.3}$$

converges for each $m = 0, 1, 2, \dots$ to a value t_m, and, second,

$$\lim_{m\to\infty} t_m = t. \tag{1.4}$$

We call the numbers t_m the *T-means* of (1.2) corresponding to the given method T.

We call the method T *row-finite* if, for each $m = 0, 1, 2, \dots,$ we have $c_{mk} = 0$ starting from some index k, which in general depends on m.

Now, given a trigonometric series

$$\sum_{j=1}^{\infty} (a_j \cos jx + b_j \sin jx), \tag{1.5}$$

we shall consider the limits of indeterminacy in measure of the T-means of (1.5) corresponding to an arbitrary method T, where we suppose that the first term of the series is $a_0/2 = 0$.

First we recall the definition of the limits of indeterminacy in measure for an arbitrary sequence of measurable functions

$$f_m(x) \qquad (m = 0, 1, 2, \dots) \tag{1.6}$$

which are finite almost everywhere on an interval $[a, b]$.

DEFINITION 1. We say that a measurable function $F(x)$, defined almost everywhere on an interval $[a, b]$, is *the upper limit in measure of the sequence* (1.6) on $[a, b]$ if, first,

$$\lim_{m\to\infty} \operatorname{mes}\{x: f_m(x) > \varphi(x),\quad \varphi(x) > F(x)\} = 0, \tag{1.7}$$

1980 *Mathematics Subject Classification*. Primary 42A24.

for every measurable function $\varphi(x)$ defined almost everywhere on $[a, b]$, and, second,

$$\varlimsup_{m\to\infty} \operatorname{mes}\{x: f_m(x) > \psi(x), \quad F(x) > \psi(x)\} > 0 \tag{1.8}$$

for every measurable function $\psi(x)$, defined almost everywhere on $[a, b]$ and satisfying

$$\operatorname{mes}\{x: F(x) > \psi(x)\} > 0. \tag{1.9}$$

DEFINITION 2. We say that a measurable function $G(x)$, defined almost everywhere on $[a, b]$, is *the lower limit in measure* on $[a, b]$ of the sequence (1.6) if $-G(x)$ is the upper limit in measure on $[a, b]$ of the sequence

$$-f_m(x) \qquad (m = 0, 1, 2, \dots). \tag{1.10}$$

In these definitions the functions $F(x)$, $G(x)$, $\varphi(x)$ and $\psi(x)$ may take the value $+\infty$ or $-\infty$ on sets of positive measure.

The upper and lower limits of (1.6) in measure on $[a, b]$ will be denoted as follows:

$$F(x) = \varlimsup_{m\to\infty} (\operatorname{mes}, [a, b]) f_m(x), \tag{1.11}$$

$$G(x) = \varliminf_{m\to\infty} (\operatorname{mes}, [a, b]) f_m(x) \tag{1.12}$$

and will be called the *limits of indeterminacy in measure* of the sequence (1.6). One can show that the upper and lower limits in measure are uniquely defined up to sets of measure zero, and that if $F(x)$ and $G(x)$ are defined by (1.11) and (1.12) then

$$G(x) \leqslant F(x) \tag{1.13}$$

almost everywhere on $[a, b]$.(1)

We have the following theorem.

THEOREM A°. *For arbitrary measurable functions* $F(x)$ *and* $G(x)$ *that satisfy* (1.13) *almost everywhere on* $[-\pi, \pi]$ *and for an arbitrary regular row-finite method* T *defined by a matrix* $\|c_{nk}\|$ *with real elements, there is a trigonometric series* (1.5) *satisfying the following conditions*:

$$1^\circ. \quad \lim_{j\to\infty} a_j = 0, \qquad \lim_{j\to\infty} b_j = 0; \tag{1.14}$$

$$2^\circ. \quad F(x) = \varlimsup_{m\to\infty} (\operatorname{mes}, [-\pi, \pi]) t_m(x),$$

$$G(x) = \varliminf_{m\to\infty} (\operatorname{mes}, [-\pi, \pi]) t_m(x), \tag{1.15}$$

where

$$t_m(x) = \sum_{k=0}^{\infty} c_{mk} s_k(x) \tag{1.16}$$

and $s_0(x) = 0$, $s_k(x)$ $(k = 1, 2, \dots)$ *are the partial sums of the series* (1.5) (*i.e.* $t_m(x)$ *are the* T*-means of* (1.5) *corresponding to the given method* T).

---

(1)See [1], the definition on p. 4 (transl. p. 198) and Theorems A and C in §2. Another definition of the lower limit in measure is given in [1] on p. 4 (transl. p. 198); the equivalence of the two definitions follows from Theorem D of [1], §2. The notation (1.11) and (1.12) was introduced in [2].

This theorem was stated without proof in [3] and proved in detail in [4].

Theorem A° was generalized by Sinanjan, who proved it for series with respect to an arbitrary orthonormal system $\{\varphi_n(x)\}$ and for arbitrary regular linear summation methods (not necessarily row-finite); see [5].

In the present paper we prove a theorem which generalizes Theorem A° in a different direction. Let us introduce another definition.

DEFINITION 3. We say that a series

$$\sum_{j=0}^{\infty} u_j' \tag{1.17}$$

is a *subseries* of (1.2) if, for each $j = 0, 1, 2, \ldots,$ either $u_j' = u_j$ or $u_j' = 0$.

We can establish the following theorem.

THEOREM 1. *For every regular row-finite method T defined by a matrix $\|c_{mk}\|$ with real elements, there is a trigonometric series* (1.5) *which satisfies* (1.14) *and also the following conditions*:

a°. *No matter how we select measurable functions $F(x)$ and $G(x)$ satisfying* (1.13) *almost everywhere on $[-\pi, \pi]$, and no matter what positive integer N_0 we select, there is a subseries*

$$\sum_{j=1}^{\infty} (a_j' \cos jx + b_j' \sin jx) \tag{1.18}$$

of (1.5) *for which*

$$F(x) = \overline{\lim_{m\to\infty}}\ (\text{mes}, [-\pi, \pi])\, T_m^{(1)}(x),$$

$$G(x) = \underline{\lim_{m\to\infty}}\ (\text{mes}, [-\pi, \pi])\, T_m^{(1)}(x), \tag{1.19}$$

where

$$T_m^{(1)}(x) = \sum_{k=0}^{\infty} c_{mk} s_k'(x) \qquad (m = 0, 1, 2, \ldots), \tag{1.20}$$

$$s_0'(x) = 0, \quad s_k'(x) = \sum_{j=1}^{k} (a_j' \cos jx + b_j' \sin jx)$$

$$(k = 1, 2, \ldots) \tag{1.21}$$

and

$$a_j' = 0, \quad b_j' = 0 \quad (1 \leqslant j \leqslant N_0). \tag{1.22}$$

The following theorem is an immediate consequence of Theorem 1.

THEOREM 2. *For every regular row-finite summation method T that satisfies the hypotheses of Theorem* 1, *there is a trigonometric series* (1.5) *satisfying* (1.14) *and the following conditions*:

a. *For every measurable function $F(x)$, defined almost everywhere on $[-\pi, \pi]$, there is a subseries* (1.18) *of* (1.5) *such that the sequence*

$$T_m^{(1)}(x) \qquad (m = 0, 1, 2, \ldots) \tag{1.23}$$

converges in measure to $F(x)$ on $[-\pi, \pi]$, where $T_m^{(1)}(x)$ is defined by (1.20) *and* (1.21).

The definition of convergence in measure when the limit function equals $+\infty$ or $-\infty$ on a set of positive measure is given in [1], §1.

Theorem 2 is derived from Theorem 1 and the following lemma:

LEMMA 1. *If the elements $f_m(x)$ of the sequence* (1.6) *are measurable and finite almost everywhere on an interval* $[a, b]$ *and if*

$$\overline{\lim_{m\to\infty}} \,(\text{mes}, [a, b]) f_m(x) = \underline{\lim_{m\to\infty}} \,(\text{mes}, [a, b]) f_m(x) = F(x)$$

almost everywhere on $[a, b]$, *then the sequence* (1.6) *converges in measure to* $F(x)$ *on* $[a, b]$ (*see* [1], §3, *Theorem* 1).

In fact, if the summation method T satisfies the hypotheses of Theorem 2, there is a trigonometric series (1.5) that satisfies the conclusions of Theorem 1. But by Lemma 1, conclusion a° of Theorem 1 implies conclusion a of Theorem 2 if we take $F(x) = G(x)$ in a°, and then the series (1.5) that we have defined satisfies all the conclusions of Theorem 2, so that that theorem is proved.

§2. For the proof of Theorem 1 we quote three lemmas from the author's previous work, prove one additional lemma, and recall two definitions.

LEMMA 2. *Let there be given a positive integer* L, *a positive number* $\sigma < 1$, *a continuous function* $\varphi(x)$ *on* $[-\pi, \pi]$, *and a regular row-finite summation method* T *defined by a matrix* (1.1) *with real elements.*

Then we can find an integer $L' > L$, *a trigonometric polynomial*

$$H(x) = \sum_{j=L+1}^{L'} (a_j \cos jx + b_j \sin jx) \tag{2.1}$$

and measurable sets E *and* G_m ($L < m \leqslant L'$), *satisfying the following conditions*:

$$1) \quad \text{mes } E < \sigma,\ E \subset [-\pi, \pi]. \tag{2.2}$$

$$2) \quad \text{mes } G_m < \sigma,\ G_m \subset [-\pi, \pi] \qquad (L < m \leqslant L'). \tag{2.3}$$

3) *If* $\rho_k(x)$ ($k = 0, 1, 2, \ldots$) *and* $B_m(x, T)$ ($m = 0, 1, 2, \ldots$) *are defined by*

$$\rho_k(x) = \begin{cases} 0 & (0 \leqslant k \leqslant L), \\ \sum_{j=L+1}^{k} (a_j \cos jx + b_j \sin jx) & (L < k \leqslant L'), \\ H(x) & (k > L'), \end{cases} \tag{2.4}$$

$$B_m(x, T) = \sum_{k=0}^{\infty} c_{mk}\rho_k(x) \quad (m = 0, 1, 2, \ldots),\,(^2) \tag{2.5}$$

then

$$B_m(x, T) = \theta_m(x)\varphi(x) + \eta_m(x) \qquad (x \in [-\pi, \pi] \setminus G_m,\quad L < m \leqslant L'), \tag{2.6}$$

where

$$|\theta_m(x)| < K, \qquad |\eta_m(x)| < \sigma \qquad (x \in [-\pi, \pi] \setminus G_m,\quad L < m \leqslant L') \tag{2.7}$$

(2)If it is not stated for which x a relation is valid, it is assumed to be satisfied for all $x \in [-\pi, \pi]$.

and K is a number depending only on T and satisfying

$$\sum_{k=0}^{\infty} |c_{mk}| < K \qquad (m = 0, 1, 2, \ldots). \tag{2.8}$$

$$4) \quad |B_m(x, T) - \varphi(x)| < \sigma \qquad (x \in [-\pi, \pi] \setminus E, \quad m \geqslant L'). \tag{2.9}$$

$$5) \quad |a_j| < \sigma, |b_j| < \sigma \quad (L < j \leqslant L'). \tag{2.10}$$

6) *If* $\rho_{\theta,k}(x)$ $(k = 0, 1, 2, \ldots)$ *is obtained from* $\rho_k(x)$ *by replacing the numbers* a_j *and* b_j $(L < j \leqslant L')$ *by* $\theta_j a_j$ *and* $\theta_j b_j$ *and if* $B_{\theta,m}(x, T)$ $(n = 0, 1, 2, \ldots)$ *is obtained from* $B_m(x, T)$ *by replacing* $\rho_k(x)$ *by* $\rho_{\theta,k}(x)$, *then*

$$B_{\theta,m}(x, T) = 0 \qquad (x \in [-\pi, \pi], 0 \leqslant m \leqslant L) \tag{2.11}$$

for all θ_j $(L < j \leqslant L')$.(3)

LEMMA 3. *Let there be given arbitrary measurable functions* $F(x)$ *and* $G(x)$ *satisfying* (1.13) *almost everywhere on an interval* $[a, b]$. *Then there are a sequence of continuous functions*

$$Q_l(x) \quad (l = 0, 1, 2, \ldots) \tag{2.12}$$

on $[a, b]$ *and two increasing sequences of positive integers* l_μ *and* λ_μ, $\mu = 0, 1, 2, \ldots$, *which satisfy the following conditions*:

$$1^\circ \quad \overline{\lim_{l\to\infty}}\; Q_l(x) = F(x), \qquad \underline{\lim_{l\to\infty}}\; Q_l(x) = G(x)$$

almost everywhere on $[a, b]$;

$$2^\circ \quad \lim_{l\to\infty} \max_{x\in[a,b]} |Q_l(x) - Q_{l-1}(x)| = 0;$$

$$3^\circ \quad \lim_{\mu\to\infty} Q_{l_\mu}(x) = F(x), \qquad \lim_{\mu\to\infty} Q_{\lambda_\mu}(x) = G(x)$$

almost everywhere on $[a, b]$.(4)

DEFINITION 4. Let $A_0 = \{[\varphi(x), \psi(x)]\}$ be a set of pairs of measurable functions, each of which is defined almost everywhere on an interval $[a, b]$. We call a pair of functions $[f(x), g(x)]$, defined almost everywhere on $[a, b]$, a *limit pair for the set* A_0 *in the sense of convergence almost everywhere on* $[a, b]$ if there is a sequence of pairs of functions $[\varphi_m(x), \psi_m(x)] \in A_0$ $(m = 0, 1, 2, \ldots)$ such that $\lim_{m\to\infty} \varphi_m(x) = f(x)$ and $\lim_{m\to\infty} \psi_m(x) = g(x)$ almost everywhere on $[a, b]$. (The pairs $[\varphi_m(x), \psi_m(x)]$ are not necessarily different for different values of m.)

DEFINITION 5. Let $A_0 = \{[\varphi(x), \psi(x)]\}$ have the same meaning as in Definition 4. We call a subset A_0' of A_0 *everywhere dense in* A_0 *in the sense of convergence almost everywhere on* $[a, b]$ if every pair $[f(x), g(x)] \in A_0$ is a limit pair for A_0' in the sense of convergence almost everywhere on $[a, b]$. (Definitions 4 and 5 were given in [6], §3 (Definitions 14 and 15).)

LEMMA 4. *Let* $A_0 = \{[\varphi(x), \psi(x)]\}$ *be a set of pairs of measurable functions* $[\varphi(x), \psi(x)]$ *defined almost everywhere on* $[a, b]$. *Then we can find a finite or countable set of pairs of functions contained in* A_0 *and everywhere dense in this set in the sense of convergence almost everywhere on* $[a, b]$.(5)

(3)[4], §2, Lemma 1.
(4)[4], §3, Lemma 2.
(5)[5], §3, Lemma 2.

We prove the following lemma as a corollary of Lemma 4.

LEMMA 5. *Let* $A_0 = \{[F(x), G(x)]\}$ *be the set of pairs of measurable functions that satisfy* (1.13) *almost everywhere on an interval* $[a, b]$. *Then we can define a finite or countable set* A_0' *of pairs of measurable functions*

$$[F_r(x), G_r(x)] \qquad (r = 0, 1, 2, \dots), \tag{2.13}$$

that satisfy the following conditions:

1°. A_0' *is everywhere dense in* A_0 *in the sense of convergence almost everywhere on* $[a, b]$.

$$2°. \quad A_0' \subset A_0. \tag{2.14}$$

3°. *The elements of the pairs* (2.13) *are finite almost everywhere on* $[a, b]$.

PROOF. On the basis of Lemma 4 we can define a finite or countable set A_1 of pairs of functions that is a subset of A_0 and is everywhere dense in that set in the sense of convergence almost everywhere on $[a, b]$. By repeating each pair in A_1 a countable number of times, we may suppose that A_1 is countably infinite.

Let

$$[\varphi_r(x), \psi_r(x)] \qquad (r = 0, 1, 2, \dots) \tag{2.15}$$

be the pairs of functions of which A_1 is composed. Since $A_1 \subset A_0$ and since, by the hypothesis of Lemma 5, the pairs of functions that belong to A_0 are measurable and satisfy (1.13) almost everywhere on $[a, b]$, the functions

$$\varphi_r(x), \quad \psi_r(x) \qquad (r = 0, 1, 2, \dots) \tag{2.16}$$

are also measurable and satisfy

$$\psi_r(x) \leqslant \varphi_r(x) \qquad (r = 0, 1, 2, \dots) \tag{2.17}$$

almost everywhere on $[a, b]$.

In addition, since A_1 is everywhere dense in A_0 in the sense of convergence almost everywhere on $[a, b]$, by Definitions 4 and 5 of §2 we can define for each pair

$$[F(x), G(x)] \in A_0 \tag{2.18}$$

a sequence of positive integers

$$r_\tau \qquad (\tau = 0, 1, 2, \dots), \tag{2.19}$$

that satisfy

$$\lim_{\tau \to \infty} r_\tau = +\infty, \tag{2.20}$$

$$\lim_{\tau \to \infty} \varphi_{r_\tau}(x) = F(x), \qquad \lim_{\tau \to \infty} \psi_{r_\tau}(x) = G(x) \tag{2.21}$$

almost everywhere on $[a, b]$.(6)

Now put, for each $r = 0, 1, 2, \dots,$

$$\varphi_r^{(0)}(x) = \begin{cases} \varphi_r(x) & \text{if } -\infty < \varphi_r(x) < +\infty, \\ r & \text{if } \varphi_r(x) = +\infty, \\ -r & \text{if } \varphi_r(x) = -\infty, \end{cases} \tag{2.22}$$

(6)In (2.21) we do not exclude the possibility that $\varphi_{r_\tau}(x)$ or $\psi_{r_\tau}(x)$ takes the values $+\infty$ or $-\infty$.

$$\psi_r^{(0)}(x) = \begin{cases} \psi_r(x) & \text{if } -\infty < \psi_r(x) < +\infty, \\ r & \text{if } \psi_r(x) = +\infty, \\ -r & \text{if } \psi_r(x) = -\infty, \end{cases} \tag{2.23}$$

$$F_r(x) = \max[\varphi_r^{(0)}(x), \psi_r^{(0)}(x)],$$
$$G_r(x) = \min[\varphi_r^{(0)}(x), \psi_r^{(0)}(x)]$$
$$(r = 0, 1, 2, \dots). \tag{2.24}$$

In addition, let E denote the set of points $x \in [a, b]$ at which all the functions (2.16) are defined and satisfy (2.17). Since the functions (2.16) are measurable and satisfy (2.17) almost everywhere on $[a, b]$, it follows, first, that

$$\text{mes } E = b - a, \qquad E \subset [a, b], \tag{2.25}$$

and second, that, by (2.22)–(2.24), the functions $\varphi_r^{(0)}(x)$, $\psi_r^{(0)}(x)$, $F_r(x)$ and $G_r(x)$ are measurable and finite for all x and r such that

$$x \in E, \qquad r = 0, 1, 2, \dots. \tag{2.26}$$

Consequently these four functions are measurable and finite almost everywhere on $[a, b]$, and hence $F_r(x)$ and $G_r(x)$ $(r = 0, 1, 2, \dots)$ satisfy condition 3° of Lemma 5. In addition, by (2.24),

$$G_r(x) \leqslant F_r(x) \qquad (r = 0, 1, 2, \dots) \tag{2.27}$$

almost everywhere on $[a, b]$. Therefore if A_0' is the set of pairs (2.13) of the functions $F_r(x)$ and $G_r(x)$ that we have defined, it will follow from the definition of A_0 (see the statement of Lemma 5) and (2.27) that A_0' satisfies condition 2° of Lemma 5.

Consequently to complete the proof of Lemma 5 we have only to show that the set A_0' satisfies condition 1° of the lemma.

Since (2.17) is satisfied for all x and r satisfying (2.26), we have, for each such x and r, one of the following inequalities:

$$1) \quad -\infty < \psi_r(x) \leqslant \varphi_r(x) < +\infty,$$
$$2) \quad -\infty = \psi_r(x) < \varphi_r(x) < +\infty,$$
$$3) \quad -\infty = \psi_r(x) = \varphi_r(x),$$
$$4) \quad \psi_r(x) = \varphi_r(x) = +\infty,$$
$$5) \quad -\infty < \psi_r(x) < \varphi_r(x) = +\infty,$$
$$6) \quad -\infty = \psi_r(x) < \varphi_r(x) = +\infty. \tag{2.28}$$

By using (2.22)–(2.24) we can then, for the x and r in question, express $F_r(x)$ and $G_r(x)$ in terms of $\varphi_r(x)$ and $\psi_r(x)$; in fact, we have

$$F_r(x) = \max[\varphi_r(x), \psi_r(x)] = \varphi_r(x),$$
$$G_r(x) = \min[\varphi_r(x), \psi_r(x)] = \psi_r(x), \tag{2.29}$$

if 1) is satisfied;

$$F_r(x) = \max[\varphi_r(x), -r],$$
$$G_r(x) = \min[\varphi_r(x), -r] \leqslant -r, \tag{2.30}$$

if 2) is satisfied;

$$F_r(x) = G_r(x) = -r, \tag{2.31}$$

if 3) is satisfied;

$$F_r(x) = G_r(x) = r, \tag{2.32}$$

if 4) is satisfied;

$$F_r(x) = \max[r, \psi_r(x)] \geqslant r,$$
$$G_r(x) = \min[r, \psi_r(x)], \tag{2.33}$$

if 5) is satisfied;

$$F_r(x) = r, \qquad G_r(x) = -r, \tag{2.34}$$

if 6) is satisfied.

Let $[F(x), G(x)]$ be any pair satisfying (2.18). Then, as we have already seen, we can find a sequence of integers (2.19) satisfying (2.20) and (2.21). Let E' be the set of points $x \in E$ for which the equalities (2.21) hold. Since these equalities hold almost everywhere on $[a, b]$, by (2.25) we have

$$\text{mes } E' = b - a, \qquad E' \subset E \subset [a, b]. \tag{2.35}$$

By the definition of E the functions $\varphi_r(x)$ and $\psi_r(x)$ $(r = 0, 1, 2, \ldots)$ are defined for all $x \in E'$ and satisfy (2.17).

Now let

$$\{\nu_s^{(j)}\} \equiv \{\nu_s^{(j)}\}_{s=0}^{s=\infty} \qquad (j = 1, \ldots, 6) \tag{2.36}$$

be the increasing sequence of nonnegative integers τ for which condition j) holds for $r = r_\tau$ and the given $x \in E'$ (see (2.28)). The sequence (2.36) depends, in general, on x, and since the conditions (2.28) are mutually exclusive and exhaustive, the six sequences are mutually disjoint and collectively exhaust the sequence $\tau = 0, 1, 2, \ldots$. Consequently at least one of the six sequences (2.36) is infinite.

Select a point $x \in E'$ and suppose that $\tau \to \infty$ through the elements of one of the sequences (2.36) corresponding to a particular value $j = 1, \ldots, 6$. Then for all these values of τ condition j) of (2.28) is satisfied with the same j for $r = r_\tau$ and the specified $x \in E'$.

Suppose first that

$$\tau \in \{\nu_s^{(1)}\}. \tag{2.37}$$

For these values of τ, condition 1) is satisfied with $r = r_\tau$, and consequently by (2.29)

$$F_{r_\tau}(x) = \varphi_{r_\tau}(x), \qquad G_{r_\tau}(x) = \psi_{r_\tau}(x) \qquad (\tau \in \{\nu_s^{(1)}\}). \tag{2.38}$$

Suppose next that

$$\tau \in \{\nu_s^{(2)}\}. \tag{2.39}$$

For these values of τ, condition 2) is satisfied with $r = r_\tau$, and consequently by (2.30)

$$F_{r_\tau}(x) = \max[\varphi_{r_\tau}(x), -r_\tau],$$
$$\psi_{r_\tau}(x) = -\infty, \qquad G_{r_\tau}(x) \leqslant -r_\tau \qquad (\tau \in \{\nu_s^{(2)}\}). \tag{2.40}$$

If now

$$\tau \in \{\nu_s^{(3)}\}, \tag{2.41}$$

then condition 3) is satisfied for $r = r_\tau$, and consequently by (2.31)

$$\varphi_{r_\tau}(x) = \psi_{r_\tau}(x) = -\infty, \qquad F_{r_\tau}(x) = G_{r_\tau}(x) = -r_\tau \qquad (\tau \in \{\nu_s^{(3)}\}). \tag{2.42}$$

Next, if

$$\tau \in \{\nu_s^{(4)}\}, \tag{2.43}$$

condition 4) is satisfied with $r = r_\tau$, and consequently by (2.32)

$$\varphi_{r_\tau}(x) = \psi_{r_\tau}(x) = +\infty, \qquad F_{r_\tau}(x) = G_{r_\tau}(x) = r_\tau \qquad (\tau \in \{\nu_s^{(4)}\}). \tag{2.44}$$

Suppose now that

$$\tau \in \{\nu_s^{(5)}\}. \tag{2.45}$$

Then condition 5) is satisfied with $r = r_\tau$, and consequently by (2.33)

$$\varphi_{r_\tau}(x) = +\infty, \qquad F_{r_\tau}(x) \geqslant r_\tau,$$
$$G_{r_\tau}(x) = \min[r_\tau, \psi_{r_\tau}(x)] \qquad (\tau \in \{\nu_s^{(5)}\}). \tag{2.46}$$

Finally suppose that

$$\tau \in \{\nu_s^{(6)}\}. \tag{2.47}$$

Then condition 6) is satisfied with $r = r_\tau$, and then by (2.34)

$$\varphi_{r_\tau}(x) = +\infty, \quad \psi_{r_\tau}(x) = -\infty, \quad F_{r_\tau}(x) = r_\tau, \quad G_{r_\tau}(x) = -r_\tau$$
$$(\tau \in \{\nu_s^{(6)}\}). \tag{2.48}$$

Since we always suppose that $x \in E'$, it follows from the definition of E' that (2.21) is satisfied for the x's under consideration. Furthermore, the numbers r_τ satisfy (2.20).

Suppose first that $\tau \to +\infty$ through values satisfying (2.37). Then (2.38) holds and consequently, by (2.21),

$$F_{r_\tau}(x) \to F(x), \qquad G_{r_\tau}(x) \to G(x). \tag{2.49}$$

Next suppose that $\tau \to +\infty$ through values satisfying (2.39). Then (2.40) holds. Let us show that (2.49) is valid in this case also. In fact, it follows from (2.20), (2.21) and (2.40) that

$$G_{r_\tau}(x) \to -\infty = G(x). \tag{2.50}$$

In addition, if $F(x) > -\infty$ for a value of x under consideration, then it follows from the same relations that for sufficiently large τ we have

$$-r_\tau < \varphi_{r_\tau}(x) = F_{r_\tau}(x) \to F(x). \tag{2.51}$$

If, however, $F(x) = -\infty$, we find from the same relations that

$$F_{r_\tau}(x) = \max[\varphi_{r_\tau}(x), -r_\tau] \to -\infty = F(x). \tag{2.52}$$

We see from (2.50)–(2.52) that (2.49) holds when $\tau \to +\infty$ through the sequence $\{\nu_s^{(2)}\}$.

Suppose next that $\tau \to +\infty$ through values satisfying (2.41). Then (2.42) is satisfied and consequently, by (2.20) and (2.21),

$$F_{r_\tau}(x) \to -\infty = F(x), \qquad G_{r_\tau}(x) \to -\infty = G(x),$$

i.e. we again obtain (2.49).

Now let $\tau \to +\infty$ through values satisfying (2.43). Then (2.44) is satisfied and consequently, by (2.20) and (2.21),

$$F_{r_\tau}(x) \to +\infty = F(x), \qquad G_{r_\tau}(x) \to +\infty = G(x),$$

and we again obtain (2.49).

Next, let $\tau \to +\infty$ through values satisfying (2.45). Then (2.46) is satisfied. Let us show that (2.49) also holds in this case. In fact, it follows from (2.20), (2.21) and (2.46) that

$$F_{r_\tau}(x) \to +\infty = F(x). \tag{2.53}$$

In addition, if $G(x) < +\infty$ for the value of x under consideration, it follows from the same relations that, for sufficiently large τ,

$$r_\tau > \psi_{r_\tau}(x) = G_{r_\tau}(x) \to G(x). \tag{2.54}$$

If, on the other hand, $G(x) = +\infty$, we obtain from the same relations that

$$G_{r_\tau}(x) = \min\left[r_\tau, \psi_{r_\tau}(x)\right] \to +\infty = G(x). \tag{2.55}$$

We see from (2.53)–(2.55) that (2.49) holds when $\tau \to +\infty$ through the sequence $\{\nu_s^{(5)}\}$.

Suppose, finally, that $\tau \to +\infty$ through values satisfying (2.47). Then (2.48) is satisfied and consequently, by (2.20) and (2.21),

$$F_{r_\tau}(x) \to +\infty = F(x), \qquad G_{r_\tau}(x) \to -\infty = G(x),$$

and we again obtain (2.49).

Hence (2.49) holds when $\tau \to +\infty$ through values in any of the six sequences (2.36). Since these sequences collectively exhaust the sequence $\tau = 0, 1, 2, \ldots$ (see the remark after the definition of (2.36)), we have

$$\lim_{\tau \to \infty} F_{r_\tau}(x) = F(x), \qquad \lim_{\tau \to \infty} G_{r_\tau}(x) = G(x) \tag{2.56}$$

for all values of x under consideration, i.e. for all $x \in E'$.

Hence it follows from (2.35) that (2.56) is satisfied almost everywhere on $[a, b]$. Since the set A_0' consists of all the pairs (2.13) (see the definition of A_0' after (2.27)), we have

$$\left[F_{r_\tau}(x), G_{r_\tau}(x)\right] \in A_0' \qquad (\tau = 0, 1, 2, \ldots),$$

and therefore by Definition 4 (§2) and (2.56) the pair $[F(x), G(x)]$ is a limit for the set A_0' in the sense of convergence almost everywhere on $[a, b]$. But we assumed that $[F(x), G(x)]$ was an arbitrary pair of measurable functions satisfying (2.18) (see the remark after (2.34)). Hence by Definition 5 of §2 and the inclusion (2.14), which has already been established, the set A_0' is dense in A_0 in the sense of convergence almost everywhere on $[a, b]$, i.e. the set A_0' satisfies condition 1° of Lemma 5, and the proof of the lemma is complete (see the remark preceding (2.28)).

§3. PROOF OF THEOREM 1, stated in §1. In this section we define a trigonometric series (1.5) that satisfies (1.14), and in the next two sections we shall show that it satisfies all the requirements of Theorem 1.

Take an arbitrary linear summation method T defined by a matrix (1.1) with real elements and satisfying the hypotheses of Theorem 1. Then the method is regular, and consequently, according to well-known theorems, satisfies (2.8), where K depends only on the summation method, and also satisfies

$$\lim_{m\to\infty} c_{mk} = 0 \qquad (k = 0, 1, 2, \ldots), \tag{3.1}$$

$$\lim_{m\to\infty} \sum_{k=0}^{\infty} c_{mk} = 1 \tag{3.2}$$

(see, for example, [6], Chapter III, Theorem 2).

Moreover, by the hypotheses of Theorem 1 the method T is row-finite; consequently

$$c_{mk} = 0 \qquad (k \geqslant k_m,\ m = 0, 1, 2, \ldots) \tag{3.3}$$

(see the definition after (1.4)), where k_m is a positive integer.

Now let A_0 be the set of pairs $[F(x), G(x)]$ of measurable functions that satisfy (1.13) almost everywhere on $[-\pi, \pi]$. Then by Lemma 5 of §2 there is a countable set A_0' of pairs (2.13) of functions that satisfy conditions 1°–3° of that lemma, with $[a, b] = [-\pi, \pi]$. Moreover, by duplicating each pair belonging to A_0', we may always suppose that

$$F_{r-1}(x) = F_r(x), \quad G_{r-1}(x) = G_r(x) \quad (r = 2, 4, 6, \ldots) \tag{3.4}$$

almost everywhere on $[-\pi, \pi]$. In addition, since (1.13) holds, by the definition of A_0, for all pairs $F(x)$ and $G(x)$ satisfying

$$[F(x), G(x)] \in A_0, \tag{3.5}$$

it follows from the inclusion (2.14) (see condition 2 of Lemma 5) that

$$G_r(x) \leqslant F_r(x) \qquad (r = 1, 2, \ldots) \tag{3.6}$$

almost everywhere on $[-\pi, \pi]$.

It follows from (3.6) and Lemma 3 of §2 that for each $r = 0, 1, 2, \ldots$ we can define a sequence

$$Q_{r,l}(x) \qquad (l = 0, 1, 2, \ldots) \tag{3.7}$$

of continuous functions on $[-\pi, \pi]$, and two increasing sequences

$$l_{r,\mu} \quad (\mu = 0, 1, 2, \ldots), \qquad \lambda_{r,\mu} \quad (\mu = 0, 1, 2, \ldots) \tag{3.8}$$

of integers, such that the conditions

$$\alpha. \quad \overline{\lim_{l\to\infty}}\, Q_{r,l}(x) = F_r(x), \quad \underline{\lim_{l\to\infty}}\, Q_{r,l}(x) = G_r(x) \qquad (r = 0, 1, 2, \ldots) \tag{3.9}$$

are satisfied almost everywhere on $[-\pi, \pi]$;

$$\beta. \quad \lim_{l\to\infty} \omega_{r,l} = 0 \qquad (r = 1, 2, \ldots), \tag{3.10}$$

where

$$\omega_{r,l} = \max_{x\in[-\pi,\pi]} |Q_{r,l}(x) - Q_{r,l-1}(x)| \qquad (r = 0, 1, 2, \ldots, l = 0, 1, 2, \ldots); \tag{3.11}$$

$$\gamma. \quad \lim_{\mu\to\infty} Q_{r,l_{r,\mu}}(x) = F_r(x), \quad \lim_{\mu\to\infty} Q_{r,\lambda_{r,\mu}}(x) = G_r(x) \qquad (r = 0, 1, 2, \ldots) \tag{3.12}$$

almost everywhere on $[-\pi, \pi]$.

By selecting, if necessary, sufficiently sparse subsequences of (3.8), we may suppose that

$$\mu < l_{r,\mu} + 1 < \lambda_{r,\mu} < l_{r,\mu+1} \qquad (\mu = 0, 1, 2, \ldots, r = 0, 1, 2, \ldots), \tag{3.13}$$

whence it follows that

$$l_{r,\mu+1} > l_{r,\mu} + 2, \qquad \lambda_{r,\mu+1} > \lambda_{r,\mu} + 2 \qquad (\mu = 0, 1, 2, \ldots, r = 0, 1, 2, \ldots). \tag{3.14}$$

Furthermore, by taking account of (3.4) we may also suppose that

$$Q_{r-1,\mu}(x) = Q_{r,\mu}(x), \qquad l_{r-1,\mu} = l_{r,\mu}, \lambda_{r-1,\mu} = \lambda_{r,\mu} \qquad (\mu = 0, 1, 2, \ldots; \quad r = 2, 4, 6, \ldots). \tag{3.15}$$

By (3.9) we have

$$\lim_{l\to\infty} Q_{r,l}^{(1)}(x) = F_r(x), \quad \lim_{l\to\infty} Q_{r,l}^{(2)}(x) = G_r(x) \qquad (r = 0, 1, 2, \ldots) \tag{3.16}$$

almost everywhere on $[-\pi, \pi]$, where

$$Q_{r,l}^{(1)}(x) = \sup_{l'>l} Q_{r,l'}(x), \qquad Q_{r,l}^{(2)}(x) = \inf_{l'>l} Q_{r,l'}(x) \qquad (r = 0, 1, 2, \ldots, \quad l = 0, 1, 2, \ldots). \tag{3.17}$$

The functions $F_r(x)$ and $G_r(x)$, $r = 0, 1, 2, \ldots$, that we have defined satisfy all the conditions of Lemma 5 of §2 with $[a, b] = [-\pi, \pi]$, and consequently are finite almost everywhere on $[-\pi, \pi]$ (see condition 3 of Lemma 5). Hence if we take account of (3.12) and (3.16) and apply Egorov's theorem to the sequences

$$Q_{r,l_{r\mu}}(x) \quad (\mu = 0, 1, 2, \ldots), \qquad Q_{r,l}^{(1)}(x) \quad (l = 0, 1, 2, \ldots), \tag{3.18}$$

$$Q_{r,\lambda_{r\mu}}(x) \quad (\mu = 0, 1, 2, \ldots), \qquad Q_{r,l}^{(2)}(x) \quad (l = 0, 1, 2, \ldots), \tag{3.19}$$

we can define sets Ω_r $(r = 0, 1, 2, \ldots)$ satisfying

$$\operatorname{mes} \Omega_r > 2\pi - \frac{1}{2^r}, \quad \Omega_r \subset [-\pi, \pi] \qquad (r = 0, 1, 2, \ldots), \tag{3.20}$$

on which the sequence (3.18) converges uniformly to $F_r(x)$ and (3.19) converges uniformly to $G_r(x)$.

It follows from this together with (3.10) and (3.17) that we can define positive integers μ_r $(r = 0, 1, 2, \ldots)$ such that

$$\omega_{r,l} < \frac{1}{2^r} \qquad (l \geqslant \mu_r, r = 0, 1, 2, \ldots), \tag{3.21}$$

$$\mu_r > r + 1 \qquad (r = 0, 1, 2, \dots), \tag{3.22}$$

$$|Q_{r,l_{r,\mu}}(x) - F_r(x)| < \frac{1}{2^r}, \quad |Q_{r,\lambda_{r,\mu}}(x) - G_r(x)| < \frac{1}{2^r}$$
$$(x \in \Omega_r, \mu \geqslant \mu_r, r = 0, 1, 2, \dots), \tag{3.23}$$

$$G_r(x) - \frac{1}{2^r} < Q_{r,l}(x) < F_r(x) + \frac{1}{2^r}$$
$$(x \in \Omega_r, l \geqslant \mu_r, r = 0, 1, 2, \dots). \tag{3.24}$$

Here we may suppose, by (3.4) and (3.15), that

$$\mu_{r-1} = \mu_r \qquad (r = 2, 4, 6, \dots), \tag{3.25}$$

$$\mu_{r-1} \leqslant \mu_r \qquad (r = 1, 2, \dots), \tag{3.26}$$

$$\lim_{r\to\infty} \mu_r = 0. \tag{3.27}$$

In addition, it follows from (3.22) that

$$\mu_r > 1 \qquad (r = 0, 1, 2, \dots), \tag{3.28}$$

and, since (3.8) increases, that

$$l_{r,\mu} \geqslant \mu, \lambda_{r,\mu} \geqslant \mu \qquad (r = 0, 1, 2, \dots, \mu = 0, 1, 2, \dots), \tag{3.29}$$

$$\lim_{\mu\to\infty} l_{r,\mu} = +\infty, \quad \lim_{\mu\to\infty} \lambda_{r,\mu} = +\infty. \tag{3.30}$$

We put

$$\nu(r) = \mu_r + 2^{r-2} \qquad (r = 2, 3, \dots), \tag{3.31}$$

$$\Pi_1 = 1, \Pi_r = 1 + \sum_{\rho=2}^{r} (l_{\rho,\nu(\rho)} - l_{\rho,\mu_\rho}) \qquad (r = 2, 3, 4, \dots), \tag{3.32}$$

whence, by (3.14) and (3.31),

$$\Pi_r - \Pi_{r-1} = l_{r,\nu(r)} - l_{r,\mu_r} > 2 \qquad (r = 2, 3, \dots), \tag{3.33}$$

$$\Pi_r \geqslant 1 \qquad (r = 1, 2, \dots). \tag{3.34}$$

In addition, we put

$$l(r, v) = l_{r,\mu_r+\nu} \quad (r = 0, 1, 2, \dots; \quad v = 0, 1, 2, \dots), \tag{3.35}$$

$$\kappa(t) = l(r, 0) + t - \Pi_{r-1} \quad (\Pi_{r-1} < t \leqslant \Pi_r, r = 2, 3, \dots), \tag{3.36}$$

$$\tilde{Q}_t(x) = Q_{r,\kappa(t)}(x) \quad (\Pi_{r-1} < t \leqslant \Pi_r, r = 2, 3, \dots). \tag{3.37}$$

Then since l_{r,μ_r} are positive integers (see (3.8)), $\Pi_1 = 1$ (see (3.32)) and the functions (3.7) are continuous on $[-\pi, \pi]$ for all $r, l = 0, 1, 2, \dots,$ it follows that the functions $\tilde{Q}_t(x)$ are defined for $t = 2, 3, \dots$ and continuous on $[-\pi, \pi]$.

Since

$$\kappa(t) > l(r, 0) = l_{r,\mu_r} \geqslant \mu_r \quad (\Pi_{r-1} < t \leqslant \Pi_r, r = 2, 3, \dots) \tag{3.38}$$

by (3.29), (3.35) and (3.36), we have, by (3.24) and (3.37),

$$G_r(x) - \frac{1}{2^r} < \tilde{Q}_t(x) < F_r(x) + \frac{1}{2^r} \qquad (\Pi_{r-1} < t \leqslant \Pi_r, x \in \Omega_r, r = 2, 3, \dots). \tag{3.39}$$

In addition, by (3.11), (3.21) and (3.36)–(3.38),

$$|\tilde{Q}_t(x) - \tilde{Q}_{t-1}(x)| = |Q_{r,\kappa(t)}(x) - Q_{r,\kappa(t)-1}(x)| < \frac{1}{2^r}$$
$$(\Pi_{r-1} + 1 < t \leqslant \Pi_r,\ r = 2, 3, \dots). \tag{3.40}$$

Taking account of the second equation (3.15), and (3.25), (3.33) and (3.35), we shall have

$$l(r-1, 0) = l_{r-1,\mu_{r-1}} = l_{r,\mu_{r-1}} = l_{r,\mu_r} = l(r, 0) \quad (r = 4, 6, 8, \dots), \tag{3.41}$$
$$l(r-1, 0) + \Pi_{r-1} - \Pi_{r-2} = l(r-1, 0) + l_{r-1,\nu(r-1)} - l_{r-1,\mu_{r-1}}$$
$$= l_{r-1,\nu(r-1)} = l_{r,\nu(r-1)} \quad (r = 4, 6, 8, \dots), \tag{3.42}$$

whence we obtain, by the first equation (3.15) and (3.36), (3.37),

$$\tilde{Q}_{\Pi_{r-1}}(x) = Q_{r-1,l_{r,\nu(r-1)}}(x) = Q_{r,l_{r,\nu(r-1)}}(x) \quad (r = 4, 6, 8, \dots), \tag{3.43}$$

and since

$$\nu(r-1) = \mu_{r-1} + 2^{r-3} = \mu_r + 2^{r-3} > \mu_r \quad (r = 4, 6, 8, \dots)$$

by (3.25) and (3.31), it follows from (3.23) and (3.43) that

$$|\tilde{Q}_{\Pi_{r-1}}(x) - F_r(x)| < \frac{1}{2^r} \quad (x \in \Omega_r,\ r = 4, 6, 8, \dots). \tag{3.44}$$

In addition, by (3.36) we have

$$\kappa(\Pi_{r-1} + 1) = l(r, 0) + 1 \quad (r = 2, 3, \dots),$$

whence, by (3.11), (3.21), (3.37) and (3.38),

$$l(r, 0) = l_{r,\mu_r} \geqslant \mu_r, \quad |\tilde{Q}_{\Pi_{r-1}+1}(x) - Q_{r,l(r,0)}| < \frac{1}{2^r} \quad (r = 2, 3, \dots),$$

and consequently we obtain

$$|\tilde{Q}_{\Pi_{r-1}+1}(x) - F_r(x)| < \frac{1}{2^{r-1}} \quad (x \in \Omega_r,\ r = 2, 3, \dots) \tag{3.45}$$

from (3.23). Then, by (3.44),

$$|\tilde{Q}_{\Pi_{r-1}+1}(x) - \tilde{Q}_{\Pi_{r-1}}(x)| < \frac{1}{2^{r-2}} \quad (x \in \Omega_r,\ r = 4, 6, 8, \dots). \tag{3.46}$$

Comparing (3.40) with (3.46), we finally obtain

$$|\tilde{Q}_t(x) - \tilde{Q}_{t-1}(x)| < \frac{1}{2^{r-2}} \quad (x \in \Omega_r,\ \Pi_{r-1} < t \leqslant \Pi_r,\ r = 4, 6, 8, \dots). \tag{3.47}$$

Since the sequence (3.8) increases, it follows from (3.29) and (3.35) that

$$l_{r,\mu_r+v} = l(r, v) \geqslant l(r, 0) = l_{r,\mu_r} \geqslant \mu_r$$
$$(r = 0, 1, 2, \dots;\quad v = 0, 1, 2, \dots), \tag{3.48}$$

whence by (3.23)

$$|Q_{r,l(r,v)}(x) - Q_{r',l(r',v')}(x)| < \frac{1}{2^{r'-1}} + |F_r(x) - F_{r'}(x)|$$
$$(x \in (\Omega_r \cap \Omega_{r'}),\ r' \leqslant r;\quad r, r' = 1, 2, \dots;\quad v, v' = 0, 1, 2, \dots). \tag{3.49}$$

In addition, put

$$t_{r,\sigma,v} = \Pi_{r-1} + l(r, \sigma + v - 1) - l(r, 0)$$

$$(\sigma = 0, 1; \quad v = 1, 2, \ldots, 2^{r-2}; \quad r = 2, 3, \ldots), \tag{3.50}$$

whence, by (3.14), (3.31), (3.33) and (3.35),

$$t_{r,0,1} = \Pi_{r-1} < t_{r,1,1} = t_{r,0,2} \leqslant t_{r,0,v} < t_{r,1,v}$$
$$= t_{r,0,v+1} < \cdots < t_{r,0,2^{r-2}+1} = t_{r,1,2^{r-2}} = \Pi_{r-1} + l_{r,\nu(r)} - l_{r,\mu_r} = \Pi_r$$
$$(r = 2, 3, \ldots; \quad 1 < v < 2^{r-2}). \tag{3.51}$$

By (3.50)

$$l(r, 0) + t_{r,\sigma,v} - \Pi_{r-1} = l(r, \sigma + v - 1)$$

$$(\sigma = 0, 1; \quad v = 1, 2, \ldots, 2^{r-2}; \quad r = 2, 3, \ldots), \tag{3.52}$$

whence, by (3.36), (3.37) and (3.51),

$$\tilde{Q}_{t_{r,\sigma,v}}(x) = Q_{r,l(r,\sigma+v-1)}(x)$$

$$(\sigma = 0, 1; \quad v = 1, 2, \ldots, 2^{r-2}; \quad \sigma + v > 1; \quad r = 2, 3, \ldots). \tag{3.53}$$

We now define inductively the sequences of positive integers

$$M_t \quad (t = 0, 1, 2, \ldots), \tag{3.54}$$

$$M_t' \quad (t = 1, 2, \ldots), \tag{3.55}$$

$$j_t \quad (t = 1, 2, \ldots), \quad v_t \quad (t = 1, 2, \ldots), \tag{3.56}$$

the real numbers, independent of t,

$$a_j, b_j \quad (j = 1, 2, \ldots), \tag{3.57}$$

and for each $t = 1, 2, \ldots$ the real numbers

$$\theta_{vj}(t) \quad (v = 1, 2, \ldots, v_t; \quad j = 0, 1, 2, \ldots, j_t). \tag{3.58}$$

Take

$$M_1 = M_1' = 1, \quad a_1 = b_1 = 0, \tag{3.59}$$

$$v_1 = 1, \qquad j_1 = 1, \quad \theta_{11}(1) = 1. \tag{3.60}$$

Suppose now that for some integer $t \geqslant 2$ we have already defined the integers

$$M_{t'}, M_{t'}', v_{t'}, j_{t'} \quad (t' = 1, 2, \ldots, t - 1) \tag{3.61}$$

and the real numbers

$$a_j, b_j \quad (j = 1, 2, \ldots, M_{t-1}), \tag{3.62}$$

$$\theta_{vj}(t') \quad (v = 1, 2, \ldots, v_{t'}; \quad j = 0, 1, 2, \ldots, j_{t'}; \quad t' = 1, 2, \ldots, t - 1). \tag{3.63}$$

Suppose also that the numbers (3.61)–(3.63) satisfy

$$M_{t'-1} < M_{t'} \qquad (2 \leqslant t' \leqslant t - 1), \tag{3.64}$$

if $t > 2$ (if $t = 2$, (3.64) is meaningless),

$$\theta_{v,j}(t') = 0 \text{ or } 1 \quad (v = 1, 2, \ldots, v_{t'}, j = 0, 1, 2, \ldots, j_{t'}, t' = 1, 2, \ldots, t - 1) \tag{3.65}$$

and

$$v_{t'} = 2^{r'-2}, \quad j_{t'} = M_{\Pi_{r'-1}} \quad (t' = 1, 2, \ldots, t - 1), \tag{3.66}$$

where $r' = r'(t')$ is determined by the inequalities

$$\Pi_{r'-1} \leqslant t' < \Pi_{r'} \quad (t' = 1, 2, \ldots, t-1). \tag{3.67}$$

Since $\Pi_1 = 1$ (see (3.32)), it follows from (3.59) and (3.60) that (3.65)–(3.67) are satisfied for $t = 2$ if we take $r' = 2$.

We now define the numbers (3.61)–(3.63), where t is replaced by $t + 1$ and j satisfies $M_{t-1} < j \leqslant M_t$. We suppose that the values of t under consideration satisfy

$$t \geqslant 2. \tag{3.68}$$

Then, since $\Pi_1 = 1$ (see (3.32)) and Π_r $(r = 2, 3, \ldots)$ satisfy (3.33), there is a unique integer $r = r(t)$ such that

$$\Pi_{r-1} < t \leqslant \Pi_r \quad (r \geqslant 2). \tag{3.69}$$

Furthermore, by (3.51) we can define a unique integer $h = h(t)$ satisfying

$$1 \leqslant h \leqslant 2^{r-2} \tag{3.70}$$

and

$$\Pi_{r-1} = t_{r,0,1} \leqslant t_{r,0,h} < t \leqslant t_{r,1,h} \leqslant t_{r,1,2^{r-2}} = \Pi_r. \tag{3.71}$$

In this section we shall always consider the integers r and h as functions of t defined by (3.69)–(3.71).

We see from (3.69) that by hypothesis the numbers (3.61) and (3.63) are already defined and satisfy (3.64)–(3.66) for $t' = \Pi_{r-1}$ and $r' = r \geqslant 2$. Consequently $j_{\Pi_{r-1}} = M_{\Pi_{r-1}}$ and the numbers $\theta_{v,j}(\Pi_{r-1})$ are defined for $v = 1, 2, \ldots, 2^{r-2}$ and $j = 0, 1, 2, \ldots, M_{\Pi_{r-1}}$. In addition, by (3.69) and (3.64), which hold for $t' = \Pi_{r-1} + 1$, we have

$$M_{\Pi_{r-1}} \leqslant M_{t-1}, \tag{3.72}$$

and then the numbers a_j and b_j (see (3.62)) have been defined for $j = 1, 2, \ldots, M_{\Pi_{r-1}}$.

It follows from the preceding discussion and (3.51) that, for a given $r = r(t)$ and all $v = 1, 2, \ldots, 2^{r-2}$, we can define the functions $H_{r,v}(x)$, $s_{r,v,k}(x)$, $B_{r,v,m}(x, T)$ and $s_{t-1,k}(x)$ by the equations

$$H_{r,v}(x) = \sum_{j=1}^{M_{\Pi_{r-1}}} \theta_{vj}(\Pi_{r-1})(a_j \cos jx + b_j \sin jx), \tag{3.73}$$

$$s_{r,v,k}(x) = \begin{cases} \sum_{j=1}^{k} \theta_{v,j}(\Pi_{r-1})(a_j \cos jx + b_j \sin jx) \\ (0 \leqslant k \leqslant M_{\Pi_{r-1}}), \\ H_{r,v}(x) \quad (M_{\Pi_{r-1}} < k), \end{cases} \tag{3.74}$$

$$B_{r,v,m}(x, T) = \sum_{k=0}^{\infty} c_{mk} s_{r,v,k}(x) \quad (m = 0, 1, 2, \ldots), \tag{3.75}$$

where c_{mk} are the elements of the matrix that defines the row-finite summation method T (see the statement of Theorem 1 in §1), and

$$s_{t-1,k}(x, v) = \begin{cases} s_{r,v,k}(x) \quad (0 \leqslant k \leqslant M(r, v)), \\ s_{r,v,M(r,v)}(x) + \sum_{j=M(r,v)+1}^{k} (a_j \cos jx + b_j \sin jx) \\ \qquad\qquad (M(r, v) < k \leqslant M_{t-1}), \\ s_{r,v,M(r,v)}(x) + \sum_{j=M(r,v)+1}^{M_{t-1}} (a_j \cos jx + b_j \sin jx) \\ \qquad\qquad (M'(r, v) < k), \end{cases} \tag{3.76}$$

where

$$M(r, v) = M_{t_{r,0,v}}, \tag{3.77}$$

$$M'(r, v) = \max[M(r, v), M_{t-1}] \tag{3.78}$$

and $t_{r,0,v}$ is defined by (3.50). Let us also put

$$B_{t-1,m}(x, v, T) = \sum_{k=0}^{\infty} c_{mk} s_{t-1,k}(x, v) \quad (m = 0, 1, 2, \ldots). \tag{3.79}$$

The sums in (3.74)–(3.78) are to be taken to be zero if the upper index is smaller than the lower index. Moreover, since (3.73)–(3.78) are written for $v = 1, 2, \ldots, 2^{r-2}$, they hold for $v = h = h(t)$ by (3.70). Then it follows from (3.78) that

$$M(r, h) \leqslant M'(r, h) = M_{t-1}, \tag{3.80}$$

since, by (3.64), (3.71) and (3.77), $M(r, h) + 1 \leqslant M_{t-1}$ if $t \neq t_{r,0,h} + 1$ and $M(r, h) = M_{t-1}$ if $t = t_{r,0,h} + 1$.

For the value of $t \geqslant 2$ under consideration, put

$$\beta_t(x) = \tilde{Q}_t(x) - B_{t-1,M_{t-1}}(x, h, T), \tag{3.81}$$

where $\tilde{Q}_t(x)$ is defined by (3.36) and (3.37), $r = r(t)$ is defined by (3.69) and $h = h(t)$ is defined by (3.70) and (3.71). Since the function (3.7) is continuous on $[-\pi, \pi]$ for $r = 0, 1, 2, \ldots,$ so is the function $\tilde{Q}_t(x)$. By hypothesis the summation method T defined by $\|c_{nk}\|$ is row-finite. Therefore by (3.74), (3.76) and (3.79) with $v = h$, and (3.81), the function $\beta_t(x)$ is also continuous on $[-\pi, \pi]$.

We now apply Lemma 2 of §2, taking

$$\varphi(x) = \beta_t(x), \quad L = M_{t-1}, \quad \sigma = \frac{1}{2^t}. \tag{3.82}$$

According to the lemma we can define an integer

$$M_t' > M_{t-1}, \tag{3.83}$$

sets

$$E_t, G_m \qquad (M_{t-1} < m \leqslant M_t') \tag{3.84}$$

and a trigonometric polynomial

$$H_t(x) = \sum_{j=M_{t-1}+1}^{M_t'} (a_j \cos jx + b_j \sin jx), \tag{3.85}$$

satisfying the following conditions:

$$1)\quad \text{mes}\, E_t < \frac{1}{2^t}, \quad E_t \subset [-\pi, \pi]. \tag{3.86}$$

$$2)\quad \text{mes}\, G_m < \frac{1}{2^t}, \quad G_m \subset [-\pi, \pi] \quad (M_{t-1} < m \leqslant M'_t). \tag{3.87}$$

3) If

$$s_k^{(t)}(x) = \begin{cases} 0 & (0 \leqslant k \leqslant M_{t-1}), \\ \sum\limits_{j=M_{t-1}+1}^{k} (a_j \cos jx + b_j \sin jx) & (M_{t-1} < k \leqslant M'_t), \\ H_t(x) & (M'_t < k), \end{cases} \tag{3.88}$$

$$B_m^{(t)}(x, T) = \sum_{k=0}^{\infty} c_{mk} s_k^{(t)}(x) \quad (m = 0, 1, 2, \dots), \tag{3.89}$$

then

$$B_m^{(t)}(x, T) = \theta_m(x)\beta_t(x) + \eta_m(x)$$
$$(x \in [-\pi, \pi] \setminus G_m, \quad M_{t-1} < m \leqslant M'_t), \tag{3.90}$$

where

$$|\theta_m(x)| < K, \quad |\eta_m(x)| < \frac{1}{2^t}$$
$$(x \in [-\pi, \pi] \setminus G_m, \quad M_{t-1} < m \leqslant M'_t) \tag{3.91}$$

and K is a number depending only on the method T and satisfying (2.8).

$$4)\quad |B_m^{(t)}(x, T) - \beta_t(x)| < \frac{1}{2^t} \qquad (x \in [-\pi, \pi] \setminus E_t, m \geqslant M'_t). \tag{3.92}$$

$$5)\quad |a_j| < \frac{1}{2^t}, |b_j| < \frac{1}{2^t} \quad (M_{t-1} < j \leqslant M'_t). \tag{3.93}$$

6) If $s_{\theta,k}^{(t)}(x)$ $(k = 0, 1, 2, \dots)$ is obtained from $s_k^{(t)}(x)$ by replacing a_j and b_j $(M_{t-1} < j \leqslant M'_t)$ by $\theta_j a_j$ and $\theta_j b_j$, and if $B_{\theta,m}^{(t)}(x, T)$ is obtained from $B_m^{(t)}(x, T)$ by replacing $s_k^{(t)}(x)$ by $s_{\theta,k}^{(t)}(x)$, then

$$B_{\theta,m}^{(t)}(x, T) = 0 \quad (x \in [-\pi, \pi], 0 \leqslant m \leqslant M_{t-1}) \tag{3.94}$$

where θ_j $(M_{t-1} < j \leqslant M'_t)$ are arbitrary constants.

Then a_j and b_j are defined for all j satisfying

$$M_{t-1} < j \leqslant M'_t. \tag{3.95}$$

In addition, by the inductive hypothesis we have already defined a_j and b_j for all j satisfying

$$0 \leqslant j \leqslant M_{t-1}. \tag{3.96}$$

Consequently a_j and b_j are defined for all j satisfying

$$1 \leqslant j \leqslant M'_t. \tag{3.97}$$

Let $\tilde{a}_j$ and $\tilde{b}_j$ $(1 \leqslant j \leqslant M'_t)$ be numbers each of which can have one of the two values

$$\tilde{a}_j = 0, a_j; \quad \tilde{b}_j = 0, b_j \quad (1 \leqslant j \leqslant M'_t), \tag{3.98}$$

respectively. We introduce the notation

$$\tilde{H}_t(x) = \sum_{j=1}^{M_t'} \left(\tilde{a}_j \cos jx + \tilde{b}_j \sin jx\right), \tag{3.99}$$

$$\tilde{s}_{t,k}(x) = \begin{cases} \sum_{j=1}^{k} \left(\tilde{a}_j \cos jx + \tilde{b}_j \sin jx\right) & (0 \leqslant k \leqslant M_t'), \\ \tilde{H}_t(x) \quad (M_t' < k), \end{cases} \tag{3.100}$$

$$\tilde{B}_{t,m}(x, T) = \sum_{k=0}^{\infty} c_{mk} \tilde{s}_{t,k}(x) \quad (m = 0, 1, 2, \dots). \tag{3.101}$$

Recall that a sum is taken to be zero if its upper index is less than its lower index. Then

$$\tilde{s}_{t,0}(x) = 0. \tag{3.102}$$

We also write

$$R_{t,m} = \sup_{\left(\tilde{a}_j, \tilde{b}_j, 1 \leqslant j \leqslant M_t', x \in [-\pi, \pi]\right)} \left|\tilde{H}_t(x) - \tilde{B}_{t,m}(x, T)\right| \tag{3.103}$$

$$(m = 0, 1, 2, \dots),$$

where sup is taken over all x satisfying $-\pi \leqslant x \leqslant \pi$, and over all $\tilde{a}_j$ and $\tilde{b}_j$ satisfying (3.98) with given a_j and b_j $(1 \leqslant j \leqslant M_t')$.

We shall show that

$$\lim_{m \to \infty} R_{t,m} = 0. \tag{3.104}$$

In fact, since the method T is regular, (3.1) and (3.2) are satisfied, and since t is given, it follows that

$$\lim_{m \to \infty} \sum_{k=0}^{M_t'} |c_{mk}| = 0, \tag{3.105}$$

$$\lim_{m \to \infty} \sum_{k=M_t'+1}^{\infty} c_{mk} = 1. \tag{3.106}$$

Taking

$$A_t = \sum_{j=1}^{M_t'} \left(|a_j| + |b_j|\right) \tag{3.107}$$

and taking account of (3.99) and (3.100), we obtain

$$|\tilde{s}_{t,k}(x)| \leqslant A_t \quad \left(x \in [-\pi, \pi], \;\; k = 0, 1, 2, \dots\right), \tag{3.108}$$

$$|\tilde{H}_t(x)| \leqslant A_t \quad \left(x \in [-\pi, \pi]\right) \tag{3.109}$$

for all $\tilde{a}_j$ and $\tilde{b}_j$ satisfying (3.98). Then since, by (3.100),

$$\sum_{k=M_t'+1}^{\infty} c_{mk} \tilde{s}_{t,k}(x) = \tilde{H}_t(x) \sum_{k=M_t'+1}^{\infty} c_{mk} \quad (m = 0, 1, 2, \dots), \tag{3.110}$$

it follows from (3.101) and (3.107)–(3.109) that

$$|\tilde{B}_{t,m}(x, t) - \tilde{H}_t(x)| \leqslant \left|\sum_{k=0}^{M'_t} c_{mk}\tilde{s}_{t,k}(x)\right| + |\tilde{H}_t(x)|\left|\sum_{k=M'_t+1}^{\infty} c_{mk} - 1\right|$$

$$\leqslant A_t\cdot\left(\sum_{k=0}^{M'_t}|c_{mk}| + \left|\sum_{k=M'_t+1}^{\infty} c_{mk} - 1\right|\right) \equiv \Gamma_{t,m} \quad (3.111)$$

$$(-\pi \leqslant x \leqslant \pi,\ m = 0, 1, 2, \dots)$$

for arbitrary $\tilde{a}_j$ and $\tilde{b}_j$ satisfying (3.98), where the $\Gamma_{t,m}$ are independent of x and of $\tilde{a}_j$ and $\tilde{b}_j$ for given a_j and b_j $(1 \leqslant j \leqslant M'_t)$.

By (3.105) and (3.106) we have

$$\lim_{m\to\infty} \Gamma_{t,m} = 0, \quad (3.112)$$

and since it follows from (3.103) and (3.111) that

$$0 \leqslant R_{t,m} \leqslant \Gamma_{t,m} \quad (m = 0, 1, 2, \dots), \quad (3.113)$$

we obtain (3.104).

Since the method T is row-finite, the functions $\tilde{B}_{t,m}(x, T)$ are trigonometric polynomials, by (3.99)–(3.102). We take

$$\Delta_{t,m',m''} = \max_{x\in[-\pi,\pi]} |\tilde{B}_{t,m'}(x, T) - \tilde{B}_{t,m''}(x, T)| \quad (m', m'' = 0, 1, 2, \dots), \quad (3.114)$$

$$\gamma_{t,m} = \sup_{\left(\tilde{a}_j,\tilde{b}_j, 1\leqslant j\leqslant M'_t, m', m'' > m\right)} \Delta_{t,m',m''}, \quad (3.115)$$

where the sup over $\tilde{a}_j$ and $\tilde{b}_j$ is taken in the same sense as in (3.103).

By (3.103),

$$|\tilde{B}_{t,m'}(x, T) - \tilde{B}_{t,m''}(x, T)| \leqslant |\tilde{B}_{t,m'}(x, T) - \tilde{H}_t(x)| + |\tilde{H}_t(x) - \tilde{B}_{t,m''}(x, T)|$$

$$\leqslant R_{t,m'} + R_{t,m''} \leqslant 2\sup_{\nu > m} R_{t,\nu} \quad (3.116)$$

$$(m', m'' > m,\ x \in [-\pi, \pi],\ m = 0, 1, 2, \dots)$$

for all $\tilde{a}_j$ and $\tilde{b}_j$ satisfying (3.98) with given a_j and b_j $(1 \leqslant j \leqslant M'_t)$. Then by (3.114) and (3.115)

$$\gamma_{t,m} \leqslant 2\sup_{\nu > m} R_{t,\nu} \quad (m = 0, 1, 2, \dots), \quad (3.117)$$

whence by (3.104)

$$\lim_{m\to\infty} \gamma_{t,m} = 0. \quad (3.118)$$

It follows from (3.118) that we can define a positive integer M_t satisfying

$$M_t > M'_t, \quad \gamma_{t,M_t} < \frac{1}{2^t}, \quad (3.119)$$

and then we obtain, by (3.115),

$$\sup_{\left(\tilde{a}_j,\tilde{b}_j, 1\leqslant j\leqslant M'_t, m', m'' > M_t\right)} \Delta_{t,m',m''} < \frac{1}{2^t}, \quad (3.120)$$

where the sup over $\tilde{a}_j$ and $\tilde{b}_j$ is taken, as before, over all $\tilde{a}_j$ and $\tilde{b}_j$ satisfying (3.98) for the given a_j and b_j $(1 \leqslant j \leqslant M_t')$.(7) Here, by (3.83) and (3.119),

$$M_{t-1} < M_t, \tag{3.121}$$

whence we obtain, by (3.64) and the equation $M_1 = 1$ (see (3.59)),

$$1 = M_1 \leqslant M'_{t'-1} < M'_{t'} \leqslant M_t \quad (2 \leqslant t' \leqslant t), \quad M_t \geqslant t. \tag{3.122}$$

We take

$$a_j = 0, \quad b_j = 0 \quad (M_t' < j \leqslant M_t). \tag{3.123}$$

Since the numbers a_j and b_j $(1 \leqslant j \leqslant M_t')$ have already been defined (see the remark after (3.96)), we have accordingly defined the numbers

$$a_j, b_j \quad (1 \leqslant j \leqslant M_t). \tag{3.124}$$

Hence if we compare (3.85) and (3.88) with (3.83), (3.119), and (3.123), we find that

$$H_t(x) = \sum_{j=M_{t-1}+1}^{M} (a_j \cos jx + b_j \sin jx), \tag{3.125}$$

$$s_k^{(t)}(x) = \begin{cases} 0 \quad (0 \leqslant k \leqslant M_{t-1}), \\ \sum\limits_{j=M_{t-1}+1}^{k} (a_j \cos jx + b_j \sin jx) \qquad (M_{t-1} < k \leqslant M_t), \\ H_t(x) \quad (M_t < k). \end{cases} \tag{3.126}$$

In addition, it follows from (3.83), (3.99), (3.119) and (3.123) that

$$\tilde{H}_t(x) = \sum_{j=1}^{M_t} \left(\tilde{a}_j \cos jx + \tilde{b}_j \sin jx\right), \tag{3.127}$$

where $\tilde{a}_j$ and $\tilde{b}_j$ are any numbers each taking one of the two values

$$\tilde{a}_j = 0, a_j; \quad \tilde{b}_j = 0, b_j \quad (1 \leqslant j \leqslant M_t) \tag{3.128}$$

respectively.

Take

$$G_m = E_t, \quad \theta_m(x) = 1, \quad \eta_m(x) = B_m^{(t)}(x, T) - \beta_t(x)$$
$$(M_t' < m \leqslant M_t, -\pi \leqslant x \leqslant \pi), \tag{3.129}$$

where $B_t(x)$ and $B_m^{(t)}(x, t)$ are defined, respectively, by (3.81) and (3.89), and E_t satisfies (3.86). Then by (3.83), (3.86), (3.87) and (3.90)–(3.92), the sets G_m and the functions $\theta_m(x)$ and $\eta_m(x)$ are defined for all m satisfying

$$M_{t-1} < m \leqslant M_t, \tag{3.130}$$

and satisfy the following conditions:

$$\operatorname{mes} G_m < \frac{1}{2^t}, \; G_m \subset [-\pi, \pi] \quad (M_{t-1} < m \leqslant M_t), \tag{3.131}$$

and

$$B_m^{(t)}(x, T) = \theta_m(x)\beta_t(x) + \eta_m(x) \quad (x \in [-\pi, \pi] \setminus G_m, M_{t-1} < m \leqslant M_t), \tag{3.132}$$

---

(7)It follows from the definition of the $\tilde{B}_{t,m}(x, T)$ (see (3.99)–(3.101)) that the only values of $\tilde{a}_j$ and $\tilde{b}_j$ that appear in these functions are those defined by (3.98).

with

$$|\theta_m(x)| < K', |\eta_m(x)| < \frac{1}{2^t} \quad (x \in [-\pi, \pi] \setminus G_m, M_{t-1} < m \leq M_t), \tag{3.133}$$

where $K' = K + 1$ depends only on the method T (see (2.8)). In addition, it follows from (3.93) and (3.123) that

$$|a_j| < \frac{1}{2^t} \quad (M_{t-1} < j \leq M_t), \tag{3.134}$$

$$|b_j| < \frac{1}{2^t} \quad (M_{t-1} < j \leq M_t). \tag{3.135}$$

Thus for the integer $t \geq 2$ under consideration we have defined the integers M_t' and M_t (see (3.83) and (3.119)) and the real numbers (3.124).

For the selected $t \geq 2$ we now define the integers v_{t,j_t}, $r'(t)$ and the numbers

$$\theta_{vj}(t) \quad (v = 1, 2, \ldots, v_t, j = 0, 1, 2, \ldots, j_t), \tag{3.136}$$

satisfying (3.65)–(3.67) with t' replaced by t, so that $r' = r'(t)$ in (3.66) and (3.67) satisfies

$$\Pi_{r'-1} \leq t \leq \Pi_{r'}. \tag{3.137}$$

For a given t we have defined an integer $r = r(t)$ by (3.69). Suppose first that

$$\Pi_{r-1} < t < \Pi_r. \tag{3.138}$$

In this case we take

$$v_t = v_{t-1} = 2^{r-2}, \quad j_t = M_{\Pi_{r-1}}, \tag{3.139}$$

$$\theta_{v,j}(t) = \theta_{v,j}(\Pi_{r-1}) \quad (v = 1, 2, \ldots, v_t, j = 0, 1, 2, \ldots, j_t), \tag{3.140}$$

where, according to (3.65) and (3.138), each $\theta_{vj}(t)$ in (3.140) is 0 or 1. It follows that the numbers v_t, j_t and $\theta_{vj}(t)$ defined by (3.139) and (3.140) satisfy (3.65) and (3.66) with t' replaced by t; and then by (3.67) and (3.138) we have

$$r' = r'(t) = r = r(t). \tag{3.141}$$

Now suppose that

$$t = \Pi_r. \tag{3.142}$$

In this case we take

$$v_t = 2^{r-1}, \quad j_t = M_{\Pi_r}. \tag{3.143}$$

It follows from (3.69) and (3.122) that

$$M_{\Pi_{r-1}} < M_t. \tag{3.144}$$

We shall consider separately the values of j that satisfy

$$0 \leq j \leq M_{\Pi_{r-1}} \tag{3.145}$$

and those that satisfy

$$M_{\Pi_{r-1}} < j \leq j_{\Pi_r} = j_t = M_{\Pi_r}. \tag{3.146}$$

In the case when (3.145) holds, we take

$$\theta_{2w-1,j}(t) = \theta_{2w,j}(t) = \theta_{w,j}(\Pi_{r-1}) \quad \left(w = 1, 2, \ldots, 2^{r-2}, 0 \leq j \leq M_{\Pi_{r-1}}\right). \tag{3.147}$$

By (3.138) the numbers on the right of (3.147) coincide with the numbers (3.63) for $t' = \Pi_{r-1}$, which we suppose to have been defined already. Then by the definition of $r(t)$ (see (3.69)) we have $r' = r'(\Pi_{r-1}) = r = r(t)$ in (3.66) and (3.67) for the t' under consideration. Since the equation

$$\theta_{u+2w-1,j}(t) = \theta_{w,j}(\Pi_{r-1}) \quad \left(u = 0, 1, \quad w = 1, 2, \dots, 2^{r-2}, 0 < j \leqslant M_{\Pi_{r-1}}\right) \tag{3.148}$$

follows from (3.147), it follows from the first equation (3.143) that the numbers $\theta_{v,j}(t)$ are defined for

$$v = 1, 2, \dots, v_t \tag{3.149}$$

and for all j satisfying (3.145). Then by the remark after (3.147) and by (3.148), (3.65) and (3.66), each $\theta_{v,j}(t)$ equals either 0 or 1 (here $t = \Pi_r$).

Now suppose that j satisfies (3.140) and (3.146). By (3.51), for

$$w = 1, 2, \dots, 2^{r-2} \tag{3.150}$$

we have the inequality

$$\Pi_{r-1} \leqslant t_{r,0,w} < t_{r,1,w} \leqslant \Pi_r. \tag{3.151}$$

On the other hand, by the equation $\Pi_1 = 1$ (see (3.32)), the inequality $r = r(t) \geqslant 2$ (see (3.69)), the first inequality (3.122) and equation (3.142), we have

$$M_{\Pi_{r-1}} \leqslant M_{t'-1} < M_{t'} \leqslant M_{\Pi_r} \quad (\Pi_{r-1} < t' \leqslant \Pi_r), \tag{3.152}$$

and consequently, by (3.142), (3.143), and (3.151), we have

$$M_{\Pi_{r-1}} \leqslant M_{t_{r,0,w}} < M_{t_{r,1,w}} \leqslant M_t = M_{\Pi_r} = j_t$$

for $w = 1, 2, \dots, 2^{r-2}$.

Supposing as before that $t = \Pi_r$, we take, for $w = 1, 2, \dots, 2^{r-2}$,

$$\theta_{2w-1,j}(t) = \theta_{2w-1,j}(\Pi_r) = \begin{cases} 1 & \left(M_{t_{r,0,w}} < j \leqslant M_{t_{r,1,w}}\right), \\ 0 & \left(M_{\Pi_{r-1}} < j \leqslant M_{t_{r,0,w}}\right), \\ 0 & \left(M_{t_{r,1,w}} < j \leqslant j_{\Pi_r} = j_t = M_{\Pi_r}\right), \end{cases} \tag{3.153}$$

$$\theta_{2w,j}(t) = \theta_{2w,j}(\Pi_r) = 0 \quad \left(M_{\Pi_{r-1}} < j \leqslant j_{\Pi_r} = j_t = M_{\Pi_r}\right). \tag{3.154}$$

Then the $\theta_{v,j}(t)$ are defined for $v = 1, 2, \dots, 2^{r-1}$ and all j satisfying (3.146), each $\theta_{v,j}(t)$ being equal to 0 or 1.

Thus if $t = \Pi_r$ (see (3.142)), it follows from (3.143) that $\theta_{v,j}(t)$ is defined and equal to 0 or 1 for $v = 1, 2, \dots, v_t$ and for all j satisfying either (3.145) or (3.146), i.e. for $j = 0, 1, 2, \dots, j_t$. Hence in the present case (3.65) is satisfied, with t' replaced by t. Furthermore,

$$r' = r'(t) = r + 1 = r(t) + 1, \quad \Pi_{r'-1} = \Pi_r = j_t \tag{3.155}$$

for the numbers $r' = r'(t)$ appearing in (3.66), by (3.67), (3.142) and the definition of $r(t)$ (see (3.69)).

We have already defined the integers v_t, j_t and the real numbers $\theta_{v,j}(t)$, also satisfying (3.65) and (3.66) with t' replaced by t, in the case when (3.138) is satisfied; and in that case we have $r' = r'(t) = r = r(t)$ (see (3.139)–(3.141) and the remark after (3.140)). Consequently, for a given t, numbers (3.58) with the required properties are defined in all cases. In addition, for the values of t under consideration, we have defined the integers M_t' and M_t (see (3.83) and (3.119)), and the real numbers (3.124). We also note that, by (3.99)–(3.101), (3.123), (3.127) and (3.128), the sup in (3.120) can be taken over all values $\tilde{a}_j$ and $\tilde{b}_j$ satisfying (3.128), not just (3.98). Then

$$\sup_{\left(\tilde{a}_j,\tilde{b}_j,1 \leqslant j \leqslant M_t,m',m'' \geqslant M_t\right)} \Delta_{t,m',m''} < \frac{1}{2^{t-1}} \qquad (t \geqslant 3), \tag{3.156}$$

where the sup is taken over the $\tilde{a}_j$ and $\tilde{b}_j$ satisfying (3.128) with arbitrary a_j and b_j $(1 \leqslant t \leqslant M_t)$. Consequently we find from (3.114) that

$$|\tilde{B}_{t,m'}(x, T) - \tilde{B}_{t,m''}(x, T)| < \frac{1}{2^t} \quad (m', m'' \geqslant M_t), \tag{3.157}$$

where the functions on the left side involve only the numbers $\tilde{a}_j$ and $\tilde{b}_j$ defined by (3.128) (see the remark on (3.120), and (3.123)).

Thus we first defined the numbers (3.59) and (3.60). Then we assumed that the integers (3.61) and the real numbers (3.62) and (3.63) were already defined and satisfied (3.64)–(3.66), where the integer $r' = r'(t')$ was defined by (3.67) and t was an arbitrary integer satisfying (3.68).([8])

Then we defined, for a given $t \geqslant 2$, the integers M_t, M_t', v_t, j_t and $r = r(t)$, and the real numbers (3.124) and (3.136), satisfying (3.65) and (3.66) with t' replaced by t, where $r' = r'(t)$ was equal either to $r = r(t)$ or to $r + 1 = r(t) + 1$, i.e. we obtained

$$v_t = 2^{r'-2}, j_t = M \qquad (r' = r \text{ or } r + 1), \tag{3.158}$$

$$\theta_{v,j}(t) = 0 \quad \text{or } 1 \qquad (v = 1, 2, \ldots, v_t); \;\; j = 0, 1, 2, \ldots, j_t. \tag{3.159}$$

In addition, we showed that (3.121) and (3.122) were satisfied. It follows from the preceding conditions and (3.152) and (3.153) that (3.64)–(3.66) are satisfied if we replace t by $t + 1$. Consequently the inductive hypotheses are satisfied if t is replaced by $t + 1$. Therefore we can define M_{t+1}, M_{t+1}', v_{t+1}, $\theta_{vj}(t + 1)$, a_j, b_j $(M_t < j \leqslant M_{t+1})$, etc.

Hence it follows that, starting from the numbers (3.59) and (3.60), we can define, step by step, the numbers (3.54)–(3.56), (3.58), and (3.124), for $t = 1, 2, \ldots$; moreover, for each j, the numbers (3.124) depend only on j. At the same time we have defined the integers $r = r(t)$ and $h = h(t)$, the functions

$$H_{r,v}(x), s_{r,v,k}(x), B_{r,v,m}(x, T), \quad s_{t-1,k}(x, v), B_{t-1,m}(x, v, T), \tag{3.160}$$

where $r = r(t)$, $v = 1, 2, \ldots, 2^{r-2}$; $k = 0, 1, 2, \ldots, m = 0, 1, 2, \ldots,$ the integers

$$M(r, v), \quad M'(r, v) \quad (r = r(t), v = 1, 2, \ldots, 2^{r-2}), \tag{3.161}$$

the functions $\beta_t(x)$ and $H_t(x)$, the sets E_t and G_m $(M_{t-1} < m \leqslant M_t)$, the functions

$$s_k^{(t)}(x), s_{\theta,k}^{(t)}(x) \quad (k = 0, 1, 2, \ldots),$$

$$B_m^{(t)}(x, T), \theta_m(x), \quad \eta_m(x), \quad B_{\theta,m}^{(t)}(x, T) \quad (M_{t-1} < m \leqslant M_t), \tag{3.162}$$

([8])We assumed (3.64) only for $t > 2$.

the numbers (3.128), the functions

$$\tilde{H}_t(x),\quad \tilde{s}_{t,k}(x)\quad (k = 0, 1, 2, \ldots),\quad \tilde{B}_{t,m}(x, T)\quad (m = 0, 1, 2, \ldots) \tag{3.163}$$

and the numbers

$$R_{t,m}, A_t, \Gamma_{t,m}, \Delta_{t,m',m''}, \gamma_{t,m}\quad (m, m', m'' = 0, 1, 2, \ldots), \tag{3.164}$$

which together with the numbers (3.54)–(3.56), (3.58) and (3.124) satisfy conditions (3.69)–(3.157), inclusive, for all t satisfying (3.68), i.e. for $t \geqslant 2$.

It follows from (3.121) that

$$\lim_{t\to\infty} M_t = \infty. \tag{3.165}$$

Then it follows from (3.124) that the numbers (3.57) are defined for $j = 0, 1, 2, \ldots$, and, as we have shown, these numbers depend only on j and not on t. Consequently we have defined the trigonometric series (1.5). We shall show in §§4 and 5 that this series satisfies the conditions of Theorem 1 as stated in §1. In doing this we shall use conditions (3.69)–(3.157), but without mentioning each time that they are valid for arbitrary $t \geqslant 2$.

§4. It follows from (3.121) that for $j = 2, 3, \ldots$ we can define t as a function of j by

$$M_{t-1} < j \leqslant M_t, \tag{4.1}$$

and that

$$\lim_{j\to\infty} t = \infty. \tag{4.2}$$

Then, by (3.134) and (3.135), (1.14) of Theorem 1 is satisfied.

We still need to show that condition a° of that theorem is satisfied.

In the present section, corresponding to arbitrary measurable functions $F(x)$ and $G(x)$ that satisfy (1.13) almost everywhere on $[-\pi, \pi]$, we shall define a subseries (1.18) of the trigonometric series (1.5) that satisfies (1.22). In the next section we shall show that this subseries satisfies all the other equations that appear in condition a° of Theorem 1 for the given $F(x)$ and $G(x)$.

At the beginning of §3 we denoted by A_0 the set of pairs $[F(x), G(x)]$ of measurable functions that satisfy (1.13) almost everywhere on $[-\pi, \pi]$. Then our selected $F(x)$ and $G(x)$ satisfy

$$[F(x), G(x)] \in A_0. \tag{4.3}$$

In addition, we defined the countable set A_0' of pairs (2.13); this set satisfies conditions 1°–3° of Lemma 5 (§2), as well as (3.4) and (3.6).

By condition 1° of Lemma 5, Definition 5 of §2, and the inclusion (4.3), our selected pair of functions $F(x)$ and $G(x)$ is a limit pair in the sense of convergence almost everywhere on $[-\pi, \pi]$ for the set A_0', and then by Definition 4 of §2 there is an increasing sequence of integers

$$r_\tau\quad (\tau = 0, 1, 2, \ldots), \tag{4.4}$$

such that

$$\lim_{\tau\to\infty} F_{r_\tau}(x) = F(x),\quad \lim_{\tau\to\infty} G_{r_\tau}(x) = G(x) \tag{4.5}$$

almost everywhere on $[-\pi, \pi]$.

By selecting, if necessary, a sufficiently rapidly increasing subsequence of (4.4), we may suppose that

$$N_0 + 1 < r_\tau, \quad r_\tau + 1 < r_{\tau+1} \quad (\tau = 0, 1, 2, \ldots), \tag{4.6}$$

where N_0 is the integer appearing in the statement of Theorem 1. We may also suppose that all the terms of (4.4) are even, since we can replace any odd r_τ by the even number $r_\tau + 1$. Then, by (3.4),

$$F_{r_\tau}(x) = F_{r_\tau+1}(x), \quad G_{r_\tau}(x) = G_{r_\tau+1}(x)$$

for all such r_τ, almost everywhere on $[-\pi, \pi]$, and consequently (4.5) will be satisfied almost everywhere on $[-\pi, \pi]$ for the modified sequence (4.4). In addition, by (4.6) the modified sequence (4.4) will still be increasing. Consequently we may assume that all the terms of the increasing sequence (4.4) are even and satisfy (4.5).

We now define a subseries (1.18) of (1.5) that satisfies condition a° of Theorem 1 for the functions $F(x)$ and $G(x)$ under consideration. We first note that, because $\Pi_1 = 1$ (see (3.32)), the number $r = r(t)$ defined by (3.69) can take all the values 2, 3, 4, . . . if t takes arbitrary values satisfying (3.68). Then, since we take $r = r(t)$ in conditions (3.69)–(3.155), we have by (3.142), (3.143) and (3.32), (3.59), (3.69)

$$v_{\Pi_r} = 2^{r-1}, \quad j_{\Pi_r} = M_{\Pi_r} \quad (r = 1, 2, \ldots). \text{(}^9\text{)} \tag{4.7}$$

Now we define a sequence of integers

$$u_r \quad (r = 1, 2, \ldots), \tag{4.8}$$

satisfying

$$1 \leqslant u_r \leqslant 2^{r-1} = v_{\Pi_r} \quad (r = 1, 2, \ldots), \tag{4.9}$$

such that for $r = 2, 3, \ldots$ the system of numbers

$$\theta_{u_r j}(\Pi_r) \quad (j = 0, 1, 2, \ldots, j_{\Pi_r}), \tag{4.10}$$

i.e. the system of numbers (3.136) with $v = u_r$ and $t = \Pi_r$, satisfies

$$\theta_{u_{r-1} j}(\Pi_{r-1}) = \theta_{u_r j}(\Pi_r) \quad (j = 0, 1, 2, \ldots, j_{\Pi_{r-1}}, \quad r = 2, 3, \ldots). \tag{4.11}$$

We note here that, by (3.33), (3.121) and (4.7),

$$M_{\Pi_{r-1}} = j_{\Pi_{r-1}} < j_{\Pi_r} = M_{\Pi_r} \quad (r = 2, 3, \ldots). \tag{4.12}$$

Take

$$u_1 = 1. \tag{4.13}$$

Then, by (3.32), (3.59) and (3.60), the system of numbers (4.10) consists, for $r = 1$, of the single number

$$\theta_{1,1}(1) = \theta_{1,1}(\Pi_1). \tag{4.14}$$

Here $v_1 = 1$ and consequently, by (4.13), condition (4.9) is satisfied for $r = 1$.

Suppose in addition that for some $r \geqslant 2$ we have already defined an integer u_{r-1} satisfying

$$1 \leqslant u_{r-1} \leqslant 2^{r-2}. \tag{4.15}$$

---

(9) From here through (5.4) we suppose that $r = 1, 2, \ldots$ is independent of t.

Take numbers

$$\theta_{u_{r-1},j}(\Pi_{r-1}) \quad (j = 0, 1, 2, \ldots, j_{\Pi_{r-1}}). \tag{4.16}$$

Since (see (4.7)) we have $v_{\Pi_{r-1}} = 2^{r-2}$ for $t = \Pi_{r-1}$, by (4.15) the numbers (4.16) coincide with the numbers (3.136) for $v = u_{r-1}$ and $t = \Pi_{r-1}$. (If $r = 2$ we have $v_{\Pi_{r-1}} = v_1 = 1$ and $j_{\Pi_{r-1}} = j_1 = 1$ by (3.32) and (3.60), and the numbers (4.16) coincide with the numbers (4.14).

For a given r we now define the number u_r so that (4.11) is satisfied. Put

$$u_r = 2u_{r-1} - 1, \tag{4.17}$$

if r is one of the numbers (4.4), and

$$u_r = 2u_{r-1}, \tag{4.18}$$

if r is not one of the numbers (4.4). Then (4.9) is satisfied because of (4.7) and (4.15).

We now show that, for the value u_r that we have defined, the numbers (4.10) satisfy (4.11). Since the numbers (4.10) coincide with the numbers (3.136) for $t = \Pi_r$ and $v = u_r$, we see from the definition of these numbers when $0 \leqslant j \leqslant M_{\Pi_{r-1}}$ (see (3.148)) and from (4.15), (4.17) and (4.18) that the equation

$$\theta_{u_r,j}(\Pi_r) = \theta_{2u_{r-1}-1,j}(\Pi_r) = \theta_{2u_{r-1},j}(\Pi_r) = \theta_{u_{r-1},j}(\Pi_{r-1}) \quad (1 \leqslant j \leqslant M_{\Pi_{r-1}}) \tag{4.19}$$

is satisfied for $r \geqslant 2$. Hence it follows, by (4.12), that the numbers (4.10) satisfy (4.11).

In addition we have, from (3.153), (3.154) and (4.7), (4.15), (4.17) and (4.18),

$$\theta_{u_r,j}(\Pi_r) = \theta_{2u_{r-1},j}(\Pi_r) = 0 \quad (M_{\Pi_{r-1}} < j \leqslant j_{\Pi_r} = M_{\Pi_r}), \tag{4.20}$$

if r is not one of the numbers (4.4), and

$$\theta_{u_r,j}(\Pi_r) = \theta_{2u_{r-1}-1,j}(\Pi_r) = \begin{cases} 1 & (M_{t'_r} < j \leqslant M_{t''_r}), \\ 0 & (M_{\Pi_{r-1}} < j \leqslant M_{t'_r}), \\ 0 & (M_{t''_r} < j \leqslant j_{\Pi_r} = M_{\Pi_r}), \end{cases} \tag{4.21}$$

if r is one of the numbers (4.4), where we have put

$$t'_r = t_{r,0,u_{r-1}}, \quad t''_r = t_{r,1,u_{r-1}} \quad (r = 2, 3, \ldots), \tag{4.22}$$

Thus we have first defined u_r when $r = 1$ (see (4.13)), and in that case (4.9) coincides with (4.13). Then we supposed that an integer u_{r-1} was already defined so that (4.15) was satisfied, where r was any integer such that $r \geqslant 2$. Next we defined an integer u_r satisfying (4.9) and proved that the numbers (4.10) satisfy (4.11). Consequently we have defined the integers (4.8) inductively for $r = 1, 2, \ldots$; and for the same values of r these numbers and the numbers (4.10) satisfy (4.9) and (4.11).

By (4.12), we can write (4.11) in the form

$$\theta_{u_{r-1},j}(\Pi_{r-1}) = \theta_{u_r,j}(\Pi_r) \quad (j = 0, 1, 2, \ldots, M_{\Pi_{r-1}}, r = 2, 3, \ldots). \tag{4.23}$$

Since, by (4.9), the numbers (4.10) coincide with the numbers (3.136) for $v = u_r$ and $t = \Pi_r$, we have, by the remark after (3.149),

$$\theta_{u_r,j}(\Pi_r) = 0 \text{ or } 1 \quad (j = 0, 1, 2, \ldots, M_{\Pi_r}, r = 1, 2, \ldots). \tag{4.24}$$

We now take

$$a_0' = 0, \quad a_1' = 0, \quad b_1' = 0, \tag{4.25}$$

$$a_j' = \theta_{u_r,j}(\Pi_r)a_j, \quad b_j' = \theta_{u_r,j}(\Pi_r)b_j \quad (M_{\Pi_{r-1}} < j \leqslant M_{\Pi_r}, r = 2, 3, \dots). \tag{4.26}$$

Since $\Pi_1 = 1$ and $M_1 = 1$ (see (3.32) and (3.59)), we have by (4.25) and (4.26) defined the numbers

$$a_0', \quad a_j', \quad b_j' \quad (j = 1, 2, \dots). \tag{4.27}$$

Consequently we have defined the trigonometric series (1.18). Here, by (4.24)–(4.26), each a_j' ($j = 0, 1, 2, \dots$) equals 0 or a_j and each b_j' ($j = 1, 2, \dots$) equals 0 or b_j. Hence (1.18) is a subseries of (1.5), by Definition 3 of §1.

In addition, it follows from (4.6), (4.20), (4.25) and (4.26) that

$$a_0' = 0, \quad a_j' = 0, \quad b_j' = 0 \quad (1 \leqslant j \leqslant M_{\Pi'}), \tag{4.28}$$

where

$$\Pi' = M_{r_0 - 1}. \tag{4.29}$$

On the other hand, from (4.6) and the inequalities $M_t \geqslant t$ for $t = 2, 3, \dots$ (see (3.122)) we obtain

$$N_0 \leqslant r_0 - 1 \leqslant M_{\Pi'},$$

and then we see from (4.28) that the coefficients of (1.18) satisfy (1.22).

We have defined the series (1.18) so that it depends on the choice of the measurable functions $F(z)$ and $G(z)$ satisfying (1.13) almost everywhere on $[-\pi, \pi]$.

§5. We shall show in this section that the series (1.18), defined in §4 in terms of $F(x)$ and $G(x)$, satisfies condition a° for these functions. This will complete the proof of Theorem 1.

From (3.59), (4.11), (4.25), (4.26) and the equation $\Pi_1 = 1$ (see (3.32)) we have

$$a_0' = \theta_{u_r,0}(\Pi_r)a_0, \quad a_j' = \theta_{u_r,j}(\Pi_r)a_j, \quad b_j' = \theta_{u_r,j}(\Pi_r)b_j$$
$$(j = 1, 2, \dots, M_{\Pi_r}, \quad r = 1, 2, \dots).\ (^{10}) \tag{5.1}$$

In addition, from (4.21) and (4.26),

$$a_j' = \begin{cases} a_j & (M_{t_r'} < j \leqslant M_{t_r''}), \\ 0 & (M_{\Pi_{r-1}} < j \leqslant M_{t_r'}), \\ 0 & (M_{t_r''} < j \leqslant M_{\Pi_r}) \ (r = 2, 3, \dots), \end{cases} \tag{5.2}$$

$$b_j' = \begin{cases} b_j & (M_{t_r'} < j \leqslant M_{t_r''}), \\ 0 & (M_{\Pi_{r-1}} < j \leqslant M_{t_r'}), \\ 0 & (M_{t_r''} < j \leqslant M_{\Pi_r}) \ (r = 2, 3, \dots), \end{cases} \tag{5.3}$$

if r is one of the numbers (4.4), where t_r' and t_r'' are defined by (4.22). In addition, by (4.20) and (4.26),

$$a_j' = 0, \quad b_j' = 0 \quad (M_{\Pi_{r-1}} < j \leqslant M_{\Pi_r}, \quad r = 2, 3, \dots), \tag{5.4}$$

if r is not one of the numbers (4.4).

(^{10})We have assumed that $a_0 = 0$ (see the remark after (1.5)).

From here on, as for (4.7), we suppose that the integer $r = r(t)$ is the function of $t = 2, 3, \ldots$ defined by (3.69). As we have already observed, $r = r(t)$ can take any of the values $2, 3, 4, \ldots$ when t assumes these values (see the remark before (4.7)). We also note that, by (3.69), $r = r(t)$ does not decrease as t increases, so that

$$r(t) \leqslant r(t+1) \qquad (t = 2, 3, \ldots), \tag{5.5}$$

if r is not one of the numbers (4.4).

From here on, as for (4.7), we suppose that the integer $r = r(t)$ is the function of $t = 2, 3, \ldots$ defined by (3.69). As we have already observed, $r = r(t)$ can take any of the values $2, 3, 4, \ldots$ when t assumes these values (see the remark before (4.7)). We also note that, by (3.69), $r = r(t)$ does not decrease as t increases, so that

$$r(t) \leqslant r(t+1) \qquad (t = 2, 3, \ldots). \tag{5.5}$$

In §3 we considered arbitrary numbers $\tilde{a}_j$ and $\tilde{b}_j$ which satisfied (3.128) for arbitrary $t = 2, 3, \ldots$. For these numbers and for $t \geqslant 2$ we considered the functions $\tilde{H}_t(x)$, $\tilde{s}_{t,k}(x)$ and $\tilde{B}_{t,m}(x, T)$ defined by (3.99)–(3.102). Then we saw that $\tilde{H}_t(x)$ also satisfied (3.127) for the same values of t.

By the statement following the definition of the numbers (4.27), the series (1.18) is a subseries of (1.5) (see Definition 3 in §1); consequently for $j = 1, 2, \ldots$ the numbers (4.27) are a special case of the numbers $\tilde{a}_j$ and $\tilde{b}_j$ $(j = 1, 2, \ldots)$. Hence we can define the functions $H'_t(x)$, $s'_{t,k}(x)$ and $B'_{t,m}(x, T)$ which are obtained from $\tilde{H}_t(x)$, $\tilde{s}_{t,k}(x)$ and $\tilde{B}_{t,m}(x, T)$ by replacing $\tilde{a}_j$ and $\tilde{b}_j$ by a'_j and b'_j. Then since, by (3.123),

$$a'_j = 0, \quad b'_j = 0 \quad (M'_t < j \leqslant M_t,\ t = 2, 3, \ldots), \tag{5.6}$$

we can replace (3.127) and (3.100)–(3.102) by

$$H'_t(x) = \sum_{j=1}^{M_t} (a'_j \cos jx + b'_j \sin jx) \qquad (t = 2, 3, \ldots), \tag{5.7}$$

$$s'_{t,0}(x) = 0,$$

$$s'_{t,k}(x) = \begin{cases} \sum_{j=1}^{k} (a'_j \cos fx + b'_j \sin jx) & (0 < k \leqslant M_t), \\ H'_t(x)(M_t < k) & (t = 2, 3, \ldots), \end{cases} \tag{5.8}$$

$$B'_{t,m}(x, T) = \sum_{k=0}^{\infty} c_{mk} s'_{t,k}(x) \quad (m = 0, 1, 2, \ldots, t = 2, 3, \ldots). \tag{5.9}$$

(Recall that a sum is taken to be zero if its upper index is smaller than its lower index.)

In addition, since (3.73) and (3.74) hold for $r = r(t)$ $(t \geqslant 2)$ and $v = 1, 2, \ldots, 2^{r-2}$, we have, by (4.15),

$$H_{r,u_{r-1}}(x) = \sum_{j=1}^{M_{\Pi_{r-1}}} \theta_{u_{r-1},j}(\Pi_{r-1})(a_j \cos jx + b_j \sin jx) \quad (r = r(t), t = 2, 3, \ldots), \tag{5.10}$$

$$s_{r,u_{r-1},k}(x) = \begin{cases} \sum_{j=1}^{k} \theta_{u_{r-1},j}(\Pi_{r-1})(a_j \cos jx + b_j \sin jx) & (0 < k \leqslant M_{\Pi_{r-1}}), \\ H_{r,u_{r-1}}(x) \quad (M_{\Pi_{r-1}} < k) & (r = r(t), t = 2, 3, \ldots), \end{cases} \tag{5.11}$$

whence by (5.1) and the inequality $r = r(t) \geqslant 2$ (see (3.69)) we obtain

$$H_{r,u_{r-1}}(x) = \sum_{j=1}^{M_{\Pi_{r-1}}} (a_j' \cos jx + b_j' \sin jx) \quad (r = r(t), t = 2, 3, \dots), \tag{5.12}$$

$$s_{r,u_{r-1},k}(x) = \begin{cases} \sum_{j=1}^{k} (a_j' \cos jx + b_j' \sin jx) & (0 \leqslant k \leqslant M_{\Pi_{r-1}}), \\ H_{r,u_{r-1}}(x) \quad (M_{\Pi_{r-1}} < k) & (r = r(t), t = 2, 3, \dots). \end{cases} \tag{5.13}$$

In addition, it follows from the definitions of $r = r(t)$ and $h = h(t)$ (see (3.69)–(3.71)) and from (4.15) and (4.22) that

$$h = h(t) = u_{r-1} \quad (t \geqslant 2, t_r' < t \leqslant t_r''), \tag{5.14}$$

and then by (3.80)

$$M(r, u_{r-1}) \leqslant M'(r, u_{r-1}) = M_{t-1} \quad (t \geqslant 2, t_r' < t \leqslant t_r''). \tag{5.15}$$

Now suppose that the integer $r = r(t)$ defined by (3.69) coincides with one of the numbers (4.4) for some $\tau = 0, 1, 2, \dots$. We shall prove that in this case

$$B_{t-1,m}'(x, T) = B_{t-1,m}(x, u_{r-1}, T) \quad (t_r' < t \leqslant t_r''; t = 3, 4, \dots; m = 0, 1, 2, \dots), \tag{5.16}$$

where $B_{t-1,m}'(x, T)$, $s_{t-1,k}'(x)$, $B_{t-1,m}(x, v, T)$, $s_{t-1,k}(x, v)$, $M(r, v)$, $M'(r, v)$, $t_{r,0,v}$, $s_{r,v,k}(x)$ and $l(r, v)$ are defined respectively by (5.9), (5.8), (3.79), (3.76)–(3.78), (3.50), (3.74) and (3.35) for an arbitrary integer $t \geqslant 2$, for arbitrary $m, k = 0, 1, 2, \dots$, and for arbitrary $v = 1, 2, \dots, 2^{r-2}$, and t_r' and t_r'' are defined by (4.22).

In proving (5.16) we first note that, by (4.15), equations (3.76)–(3.79) and (3.50) are satisfied for $v = u_{r-1}$. Then, since $r = r(t) \geqslant 2$ for $t \geqslant 2$ (see (3.68) and (3.69)), we have by (4.22)

$$s_{t-1,k}(x, u_{r-1}) = s_{r,u_{r-1},k}(x) \quad (0 \leqslant k \leqslant M_{t_r'}, t \geqslant 2). \tag{5.17}$$

On the other hand, for the value $r = r(t)$ under consideration we have, by (3.51), (4.15) and (4.22),

$$\Pi_{r-1} \leqslant t_r' < t_r'' \leqslant \Pi_r \quad (t \geqslant 2). \tag{5.18}$$

Consequently we see that if we take account of the equation $\Pi_1 = 1$ (see (3.32)), the inequality $r = r(t) \geqslant 2$ (see (3.69)), and the inequality $M_{\Pi_1} = M_1 = 1 \leqslant M_t$ for $t \geqslant 1$ (see (3.59) and (3.121)), we have

$$1 \leqslant M_{\Pi_{r-1}} \leqslant M_{t_r'} < M_{t_r''} \quad (t \geqslant 2) \tag{5.19}$$

and consequently by (5.13) and (5.17)

$$s_{t-1,k}(x, u_{r-1}) = \sum_{j=1}^{k} (a_j' \cos jx + b_j' \sin jx) \quad (0 \leqslant k \leqslant M_{\Pi_{r-1}}, t_r' < t \leqslant t_r''). \tag{5.20}$$

In the deduction of (5.16) we supposed that $r = r(t)$ coincided with one of the numbers (4.4). In addition, by (5.12) and (5.13),

$$s_{r,u_{r-1},k}(x) = \sum_{j=1}^{M_{\Pi_{r-1}}} (a_j' \cos jx + b_j' \sin jx) \quad (k > M_{\Pi_{r-1}}; t \geqslant 2), \tag{5.21}$$

and then, by (5.2), (5.3), (5.19) and the inequality $r = r(t) > 2$ (see (3.69)),

$$s_{r,u_{r-1},k}(x) = \sum_{j=1}^{k} (a_j' \cos jx + b_j' \sin jx) \quad (0 \leq k \leq M_{t_r'}, t_r' < t \leq t_r''). \tag{5.22}$$

Comparing (5.17) with (5.22), we obtain

$$s_{t-1,k}(x, u_{r-1}) = \sum_{j=1}^{k} (a_j' \cos jx + b_j' \sin jx) \quad (0 \leq k \leq M_{t_r'}, t_r' < t \leq t_r''). \tag{5.23}$$

In addition, since we assumed that $r = r(t)$ is equal to one of the numbers (4.4), and since (3.76) and (3.77) hold for $v = u_{r-1}$ (see the remark before (5.17)), we have from these equations and (4.22), (5.2), (5.3), (5.22) and (3.121)([11])

$$s_{t-1,k}(x, u_{r-1}) = \sum_{j=1}^{M_{t_r'}} (a_j' \cos jx + b_j' \sin jx) + \sum_{j=M_{t_r'}+1}^{k} (a_j \cos jx + b_j \sin jx)$$
$$= \sum_{j=1}^{k} (a_j' \cos jx + b_j' \sin jx) \quad (M_{t_r'} < k \leq M_{t-1}, t_r' + 1 < t \leq t_r''),$$

whence by (5.23)

$$s_{t-1,k}(x, u_{r-1}) = \sum_{j=1}^{k} (a_j' \cos jx + b_j' \sin jx) \quad (0 \leq k \leq M_{t-1}, t_r' < t \leq t_r''). \tag{5.24}$$

From (3.76) with $v = u_{r-1}$, (3.121), (4.22), (5.2), (5.3), (5.15) and (5.22), we also obtain

$$s_{t-1,k}(x, u_{r-1}) = \sum_{j=1}^{M_{t-1}} (a_j' \cos jx + b_j' \sin jx) \quad (M_{t-1} < k, t_r' < t \leq t_r''), \tag{5.25}$$

if $r = r(t)$ is equal to one of the numbers (4.4). In addition, by (5.7) and (5.8),

$$H_{t-1}'(x) = \sum_{j=1}^{M_{t-1}} (a_j' \cos jx + b_j' \sin jx) \quad (t = 3, 4, \ldots), \tag{5.26}$$

$$s_{t-1,k}'(x) = \begin{cases} \sum_{j=1}^{k} (a_j' \cos jx + b_j' \sin jx) & (0 \leq k \leq M_{t-1}), \\ H_{t-1}'(x) \quad (M_{t-1} < k) & (t = 3, 4, \ldots). \end{cases} \tag{5.27}$$

Comparing the last two equations with (5.24) and (5.25), we obtain

$$s_{t-1,k}'(x) = s_{t-1,k}(x, u_{r-1}) \quad (t_r' < t \leq t_r'', t = 3, 4, \ldots, k = 0, 1, 2, \ldots), \tag{5.28}$$

if $r = r(t)$ is defined by (3.69) and is equal to one of the numbers (4.4); and since (3.79) holds for $v = u_{r-1}$ (see the remark before (5.17)), we obtain (5.16) from (5.9) and (5.28) if $r = r(t)$ is defined by (3.69) and is equal to one of the numbers (4.4).

([11])As we noted at the end of §3, (3.121) is satisfied for $t > 2$.

By (3.81), in which $h = h(t)$ is defined by (3.70) and (3.71), and which is valid for integers $t \geqslant 2$, we obtain (taking account of (5.14) and (5.16))

$$\beta_t(x) = \tilde{Q}_t(x) - B'_{t-1,M_{t-1}}(x, T) \quad (t'_r < t \leqslant t''_r,\ t = 3, 4, \ldots), \tag{5.29}$$

if $r = r(t)$ is defined by (3.69) and is equal to one of the numbers (4.4), where $\tilde{Q}_t(x)$ is defined by (3.37) and (3.36).

In addition, by (3.121), (5.2), (5.3), (5.26) and (5.18),

$$s'_{t-1,k}(x) = \begin{cases} \sum_{j=1}^{k} (a'_j \cos jx + b'_j \sin jx) & (0 \leqslant k \leqslant M_{\Pi_{r-1}}), \\ \sum_{j=1}^{M_{\Pi_{r-1}}} (a'_j \cos jx + b'_j \sin jx) & (M_{\Pi_{r-1}} < k) \end{cases} \tag{5.30}$$

$$(\Pi_{r-1} < t \leqslant t'_r,\ t = 3, 4, \ldots),$$

$$s'_{t-1,k}(x) = \begin{cases} \sum_{j=1}^{k} (a'_j \cos jx + b'_j \sin jx) & (0 \leqslant k \leqslant M_{t''_r}), \\ \sum_{j=1}^{M_{t''_r}} (a'_j \cos jx + b'_j \sin jx) & (M_{t''_r} < k) \end{cases} \tag{5.31}$$

$$(t''_r < t \leqslant \Pi_r,\ t = 3, 4, \ldots),$$

if $\Pi_{r-1} < t'_r$ or $t''_r < \Pi_r$ and if $r = r(t)$ is defined by (3.69) and is equal to one of the numbers (4.4).

Now suppose that $r = r(t)$ is defined as before by (3.69), but is not equal to one of the numbers (4.4). Then since $r = r(t) \geqslant 2$ by (3.69), equation (5.4) is satisfied and consequently, by (3.121), (5.26) and (5.27),

$$s'_{t-1,k}(x) = \begin{cases} \sum_{j=1}^{k} (a'_j \cos jx + b'_j \sin jx) & (0 \leqslant k \leqslant M_{\Pi_{r-1}}), \\ \sum_{j=1}^{M_{\Pi_{r-1}}} (a'_j \cos jx + b'_j \sin jx) & (M_{\Pi_{r-1}} < k) \end{cases} \tag{5.32}$$

$$(\Pi_{r-1} < t \leqslant \Pi_r,\ t = 3, 4, \ldots),$$

if $r = r(t)$ is defined by (3.69) and is not equal to one of the numbers (4.4).

Putting

$$\rho'_\tau = t'_{r_\tau}, \quad \rho''_\tau = t''_{r_\tau} \quad (\tau = 0, 1, 2, \ldots) \tag{5.33}$$

and taking account of (3.32), (3.33), (4.6) and (5.18), we have

$$1 = \Pi_1 \leqslant \Pi_{r_{\tau-1}-1} \leqslant \rho'_{\tau-1} < \rho''_{\tau-1} \leqslant \Pi_{r_{\tau-1}} < \Pi_{r_\tau - 1} \leqslant \rho'_\tau < \rho''_\tau \leqslant \Pi_{r_\tau}$$
$$(\tau = 1, 2, \ldots), \tag{5.34}$$

and then we can define the function $\alpha = \alpha(t)$ for the integers $t > t''_{r_0} = \rho''_0$ by

$$\alpha(t) = \begin{cases} t & (\rho'_\tau < t \leqslant \rho''_\tau), \\ \rho''_{t-1} & (\rho''_{\tau-1} < t \leqslant \rho'_\tau) \end{cases} \tag{5.35}$$

$$(\tau = 1, 2, \ldots).$$

It follows from the last two relations and (5.34) that

$$\alpha(t+1) \geqslant \alpha(t) \quad (t > \rho_0''),$$
$$2 \leqslant \alpha(\rho_\tau') = \rho_{\tau-1}'' < \rho_\tau' < \rho_\tau' + 1 = \alpha(\rho_\tau' + 1) \quad (\tau = 1, 2, \ldots), \tag{5.36}$$

whence

$$\lim_{t\to\infty} \alpha(t) = +\infty. \tag{5.37}$$

Now take

$$Q_t^{(0)}(x) = \tilde{Q}_{\alpha(t)}(x) \quad (t > \rho_0''), \tag{5.38}$$
$$\tilde{\beta}_t(x) = \beta_{\alpha(t)}(x) \quad (t > \rho_0''), \tag{5.39}$$

whence by (5.35)

$$\tilde{\beta}_t(x) = \beta_t(x) \quad (\rho_\tau' < t \leqslant \rho_\tau'', \tau = 1, 2, \ldots). \tag{5.40}$$

Take, in addition,

$$\tilde{H}_t'(x) = \sum_{j=M_{t-1}+1}^{M_t} (a_j' \cos jx + b_j' \sin jx) \quad (t = 2, 3, \ldots), \tag{5.41}$$

$$\tilde{s}_k^{(t)}(x) = \begin{cases} 0 & (0 \leqslant k \leqslant M_{t-1}), \\ \sum_{j=M_{t-1}+1}^{k} (a_j' \cos jx + b_j' \sin jx) & (M_{t-1} < k \leqslant M_t), \\ \tilde{H}_t'(x) & (M_t < k) \end{cases} \tag{5.42}$$
$$(t = 2, 3, \ldots),$$

i.e. $\tilde{H}_t'(x)$ and $\tilde{s}_k^{(t)}(x)$ are obtained from (3.125) and (3.126) by replacing a_j and b_j by a_j' and b_j'. We also take

$$\tilde{B}_m^{(t)}(x, T) = \sum_{k=0}^{\infty} c_{mk} \tilde{s}_k^{(t)}(x) \quad (t = 2, 3, \ldots; m = 0, 1, 2, \ldots). \tag{5.43}$$

Since, by (5.34) and (4.6),

$$1 \leqslant \Pi_{r_\tau - 1} \leqslant \rho_\tau' < \rho_\tau'' \leqslant \Pi_{r_\tau}, r_\tau \geqslant 2 \quad (\tau = 0, 1, 2, \ldots), \tag{5.44}$$

on comparing (5.41) and (5.42) with (3.125) and (3.126) and taking account of (5.2), (5.3), (5.33) and (3.121), we shall have

$$\tilde{s}_k^{(t)}(x) = s_k^{(t)}(x) \quad (\rho_\tau' < t \leqslant \rho_\tau''; k = 0, 1, 2, \ldots; \tau = 0, 1, 2, \ldots), \tag{5.45}$$
$$\tilde{s}_k^{(t)}(x) = 0 \quad (\Pi_{r_\tau - 1} < t \leqslant \rho_\tau'; k = 0, 1, 2, \ldots; \tau = 1, 2, \ldots), \tag{5.46}$$
$$\tilde{s}_k^{(t)}(x) = 0 \quad (\rho_\tau'' < t \leqslant \Pi_{r_\tau}; k = 0, 1, 2, \ldots; \tau = 1, 2, \ldots) \tag{5.47}$$

(here $r = r(t)$ is equal to one of the numbers (4.4)). In addition, it follows from (3.69), (3.121), (5.4), (5.41) and (5.42) that

$$\tilde{s}_k^{(t)}(x) = 0 \quad (\Pi_{r-1} < t \leqslant \Pi_r; k = 0, 1, 2, \ldots), \tag{5.48}$$

if $r = r(t)$ is not equal to any of the numbers (4.4).

If we replace τ by $\tau - 1$ in (5.47) and take account of (5.34) and (5.46), we find that

$$\tilde{s}_k^{(t)}(x) = 0 \quad (\rho_{\tau-1}'' < t \leqslant \rho_\tau'; k = 0, 1, 2, \ldots; \tau = 2, 3, \ldots), \tag{5.49}$$

and since $\tau \geqslant 3$ by (5.34), it follows from (5.43) that if $t > \rho''_{\tau-1}$ and $\tau = 1, 2, \dots$ then

$$\tilde{B}_m^{(t)}(x, T) = 0 \quad (\rho''_{\tau-1} < t \leqslant \rho'_\tau;\ m = 0, 1, 2, \dots;\ \tau = 2, 3, \dots). \tag{5.50}$$

In addition, since $t \geqslant 3$ by (5.34), if we compare (3.89) and (5.43) with (5.45) when $t > \rho'_\tau$ and $\tau = 1, 2, \dots,$ we obtain

$$\tilde{B}_m^{(t)}(x, T) = B_m^{(t)}(x, T) \quad (\rho'_\tau < t \leqslant \rho''_\tau;\ \tau = 0, 1, 2, \dots;\ m = 0, 1, 2, \dots). \tag{5.51}$$

By (3.132) and (5.51)

$$\tilde{B}_m^{(t)}(x, T) = \theta_m(x)\beta_t(x) + \eta_m(x)$$
$$(x \in [-\pi, \pi] \setminus G_m,\ M_{t-1} < m \leqslant M_t) \quad (\rho'_\tau < t \leqslant \rho''_\tau;\ \tau = 0, 1, 2, \dots), \tag{5.52}$$

where G_m, $\theta_m(x)$ and $\eta_m(x)$ satisfy (3.131) and (3.133) for $t \geqslant 2$ (see the remark at the end of §3).

Now put

$$\tilde{G}_m = G_m \quad (M_{t-1} < m \leqslant M_t,\ \rho'_\tau < t \leqslant \rho''_\tau,\ \tau = 1, 2, \dots), \tag{5.53}$$
$$\tilde{G}_m = \varnothing \quad (M_{t-1} < m \leqslant M_t,\ \rho''_{\tau-1} < t \leqslant \rho'_\tau,\ \tau = 1, 2, \dots). \tag{5.54}$$

Then by (5.34) and by (3.121), which holds for $t \geqslant 2$ (see the remark at the end of §3), the sets $\tilde{G}_m$ are defined for $m > M_{\rho''_0}$. In addition, it follows from (3.131) that

$$\operatorname{mes} \tilde{G}_m < \frac{1}{2^t}, \quad \tilde{G}_m \subset [-\pi, \pi] \quad (M_{t-1} < m \leqslant M_t,\ t \geqslant \rho''_0), \tag{5.55}$$

and then

$$\lim_{m \to \infty} \operatorname{mes} \tilde{G}_m = 0. \tag{5.56}$$

In addition, put

$$\tilde{\theta}_m(x) = \theta_m(x), \quad \tilde{\eta}_m(x) = \eta_m(x) \quad (M_{t-1} < m \leqslant M_t,\ \rho'_\tau < t \leqslant \rho''_\tau,\ \tau = 1, 2, \dots),$$
$$\tilde{\theta}_m(x) = 0, \quad \tilde{\eta}_m(x) = 0 \quad (M_{t-1} < m \leqslant M_t,\ \rho''_{\tau-1} < t \leqslant \rho'_\tau,\ \tau = 1, 2, \dots). \tag{5.57}$$

Then, by (3.121), the functions $\tilde{\theta}_m(x)$ and $\tilde{\eta}_m(x)$ are defined for $m > M_{\rho''_0}$, and we have, by (3.121), (3.133), (5.53), (5.54), (5.50), (5.52) and (5.40),

$$|\tilde{\theta}_m(x)| < K', \quad |\tilde{\eta}_m(x)| < \frac{1}{2^t} \quad (x \in [-\pi, \pi] \setminus \tilde{G}_m,\ M_{t-1} < m \leqslant M_t,\ t > \rho''_0), \tag{5.58}$$

$$\tilde{B}_m^{(t)}(x, T) = \tilde{\theta}_m(x)\tilde{\beta}_t(x) + \tilde{\eta}_m(x) \quad (x \in [-\pi, \pi] \setminus \tilde{G}_m,\ M_{t-1} < m \leqslant M_t,\ t > \rho''_1), \tag{5.59}$$

where K' depends only on the method T.

Since the numbers (4.4) are even (see the remark after (4.6)) it follows from (4.6) that when $\tau = 0, 1, 2, \dots$ the number r_τ can take only the values $4, 6, 8, \dots;$ then, comparing (3.47) with (5.34), (5.35) and (5.38), we obtain

$$|Q_t^{(0)}(x) - Q_{t-1}^{(0)}(x)| < \frac{4}{2^{r_\tau}} \quad (x \in \Omega_{r_\tau},\ \rho'_\tau + 1 < t \leqslant \rho''_\tau,\ \tau = 0, 1, 2, \dots). \tag{5.60}$$

In addition, by (5.35)

$$\alpha(\rho'_\tau) = \alpha(\rho''_{\tau-1}) = \rho''_{\tau-1} \quad (\tau = 2, 3, \dots),$$
$$\alpha(\rho'_\tau + 1) = \rho'_\tau + 1 \quad (\tau = 1, 2, \dots),$$

and, by (5.34), $\rho'_\tau > \rho''_{\tau-1} > \rho'_0$ for $\tau \geqslant 2$. Hence it follows from (5.35) and (5.38) that

$$Q_t^{(0)}(x) = Q_{t-1}^{(0)}(x) = \tilde{Q}_{\rho''_{\tau-1}}(x) = Q^{(0)}_{\rho''_{\tau-1}}(x) = Q^{(0)}_{\rho'_\tau}(x)$$
$$(\rho''_{\tau-1} < t \leqslant \rho'_\tau, \tau = 2, 3, \dots), \tag{5.61}$$
$$Q^{(0)}_{\rho'_\tau+1}(x) = \tilde{Q}_{\rho'_\tau+1}(x) \quad (\tau = 2, 3, \dots). \tag{5.62}$$

Since the numbers r_τ in the sequence (4.4) are even (see the remark after (4.6)), if we take account of (3.47) with $t = \rho'_\tau + 1$, (5.44), and (5.62), we obtain

$$|Q^{(0)}_{\rho'_\tau+1}(x) - \tilde{Q}_{\rho'_\tau}(x)| < \frac{4}{2^{r_\tau}} \quad (x \in \Omega_{r_\tau}, \tau = 2, 3, \dots). \tag{5.63}$$

Put

$$u_{r_\tau-1} = u'_\tau \quad (\tau = 0, 1, 2, \dots), \tag{5.64}$$

where u_r, $r = 1, 2, \dots,$ is defined recursively by (4.13)–(4.18). Then, by (4.6) and (4.9),

$$1 \leqslant u'_\tau \leqslant 2^{r_\tau-2} \quad (\tau = 0, 1, 2, \dots). \tag{5.65}$$

In addition, taking first

$$r = r_{\tau-1}, \quad \sigma = 1, \quad v = u'_{\tau-1} \quad (\tau = 1, 2, \dots), \tag{5.66}$$

and then

$$r = r_\tau, \quad \sigma = 0, \quad v = u'_\tau \quad (\tau = 0, 1, 2, \dots) \tag{5.67}$$

and taking account of (3.35), (4.22), (5.33) and (5.64), we have

$$t_{r\sigma v} = \rho''_{\tau-1}, \quad l(r, \sigma + v - 1) = l(r_{\tau-1}, u'_{\tau-1}) \equiv l'_\tau \quad (\tau = 1, 2, \dots), \tag{5.68}$$

if r, σ, and v are determined from (5.66), but

$$t_{r\sigma v} = \rho'_\tau, \quad l(r, \sigma + v - 1) = l(r_\tau, u'_\tau - 1) \equiv l''_\tau \quad (\tau = 1, 2, \dots), \tag{5.69}$$

if r, σ, and v are determined from (5.67).

Since u'_τ satisfies (5.65), it can assume, in particular, the values 1 or $2^{r_\tau-1}$. If

$$u'_\tau = 1 \tag{5.70}$$

for some value $\tau = 1, 2, \dots,$ then by (5.33), (5.64), (4.22) and (3.51),

$$\rho'_\tau = t'_{r_\tau} = t_{r_\tau,0,u'_\tau} = t_{r_\tau,0,1} = \Pi_{r_\tau-1} = t_{r_\tau-1,1,u''_\tau}, \tag{5.71}$$

where

$$u''_\tau = 2^{r_\tau-3}, \tag{5.72}$$

whence it follows, by the remark before (5.60), that

$$1 < u''_\tau = 2^{(r_\tau-1)-2} \quad (\tau = 1, 2, \dots). \tag{5.73}$$

Then, taking

$$r = r_\tau - 1, \quad \sigma = 1, \quad v = u''_\tau \quad (\tau = 1, 2, \dots), \tag{5.74}$$

we shall have, by (5.71),

$$t_{r,\sigma,v} = \rho'_\tau, \quad l(r, \sigma + v - 1) = l(r_\tau - 1, u''_\tau) \equiv l'''_\tau, \tag{5.75}$$

if (5.70) is satisfied for some value $\tau = 1, 2, \ldots$.

We see by (5.65) and (5.73) that $\sigma + v > 1$, if σ, v, and r satisfy (5.66) with $u'_r \geqslant 1$, or (5.67) with $u'_\tau > 1$, or (5.74) with $u'_\tau = 1$. Taking account of this, we substitute the numbers σ, v and r defined by (5.66), (5.67) and (5.74) into (3.53). Then, from (5.68), (5.69) and (5.75), we obtain

$$\tilde{Q}_{\rho''_{\tau-1}}(x) = Q_{r_\tau - 1, l'_\tau}(x) \quad (\tau = 1, 2, \ldots, u'_r \geqslant 1), \tag{5.76}$$

$$\tilde{Q}_{\rho'_\tau}(x) = Q_{r_\tau, l''_\tau}(x) \quad (\tau = 1, 2, \ldots), \tag{5.77}$$

if $u'_\tau > 1$, and

$$\tilde{Q}_{\rho'_\tau}(x) = Q_{r_\tau - 1, l'''_\tau}(x) \quad (\tau = 1, 2, \ldots), \tag{5.78}$$

if $u'_\tau = 1$. (In (3.53) we take σ, v and r respectively from (5.66), (5.67), and (5.74).)

Comparing (3.35), where μ_r is defined by (3.21) and (3.22), with (5.68), (5.69), (5.75), (5.65) and (5.73), and taking

$$\mu_{r_\tau} = \mu'_\tau, \quad \mu_{r_\tau - 1} = \mu''_\tau \quad (\tau = 0, 1, 2, \ldots), \tag{5.79}$$

we find that

$$l'_\tau = l_{r_\tau - 1, \mu'_{\tau-1} + u'_{\tau-1}} \quad (\tau = 1, 2, \ldots), \tag{5.80}$$

$$l''_\tau = l_{r_\tau, \mu'_\tau + u'_\tau - 1}, \quad l'''_\tau = l_{r_\tau - 1, \mu''_\tau + u''_\tau} \quad (\tau = 1, 2, \ldots), \tag{5.81}$$

and then it follows from (3.23), (5.65), (5.73) and (5.79) that

$$|Q_{r_{\tau-1}, l'_\tau}(x) - F_{r_{\tau-1}}(x)| < \frac{1}{2^{r_{\tau-1}}} \quad (x \in \Omega_{r_{\tau-1}}, \tau = 1, 2, \ldots), \tag{5.82}$$

$$|Q_{r_\tau, l''_\tau}(x) - F_{r_\tau}(x)| < \frac{1}{2^{r_\tau}} \quad (x \in \Omega_{r_\tau}, \tau = 1, 2, \ldots), \tag{5.83}$$

$$|Q_{r_\tau - 1, l'''_\tau}(x) - F_{r_\tau - 1}(x)| < \frac{1}{2^{r_\tau - 1}} \quad (x \in \Omega_{r_\tau - 1}, \tau = 1, 2, \ldots). \tag{5.84}$$

Then, taking account of (5.76)–(5.78), we obtain

$$|\tilde{Q}_{\rho''_{\tau-1}}(x) - F_{r_{\tau-1}}(x)| < \frac{1}{2^{r_{\tau-1}}} \quad (x \in \Omega_{r_{\tau-1}}, \tau = 1, 2, \ldots, u'_\tau \geqslant 1), \tag{5.85}$$

$$|\tilde{Q}_{\rho'_\tau}(x) - F_{r_\tau}(x)| < \frac{1}{2^{r_\tau}} \quad (x \in \Omega_{r_\tau}, \tau = 1, 2, \ldots), \tag{5.86}$$

if $u'_\tau > 1$, and

$$|\tilde{Q}_{\rho'_\tau}(x) - F_{r_\tau - 1}(x)| < \frac{2}{2^{r_\tau}} \quad (x \in \Omega_{r_\tau - 1}, \tau = 1, 2, \ldots), \tag{5.87}$$

if $u'_\tau = 1$. Since the numbers (4.4) are even, the inequality

$$|\tilde{Q}_{\rho'_\tau}(x) - F_{r_\tau}(x)| < \frac{2}{2^{r_\tau}} \quad (x \in \Omega_{r_\tau - 1}, \tau = 1, 2, \ldots) \tag{5.88}$$

follows from (3.4) and (5.87) if $u'_\tau = 1$.

For every admissible value u'_τ we find from (5.85), (5.86) and (5.88) that

$$|\tilde{Q}_{\rho''_{\tau-1}}(x) - \tilde{Q}_{\rho'_\tau}(x)| \leq |\tilde{Q}_{\rho''_{\tau-1}}(x) - F_{r_{\tau-1}}(x)| + |F_{r_{\tau-1}}(x) - F_{r_\tau}(x)|$$

$$+|\tilde{Q}_{\rho'_\tau}(x) - F_{r_\tau}(x)| \leq \frac{1}{2^{r_{\tau-1}}} + \frac{2}{2^{r_\tau}} + |F_{r_{\tau-1}}(x) - F_{r_\tau}(x)|$$

$$\left(x \in (\Omega_{r_{\tau-1}} \cap \Omega_{r_\tau} \cap \Omega_{r_\tau - 1}),\quad \tau = 1, 2, \dots\right). \tag{5.89}$$

Comparing the last inequality with (5.63), we obtain

$$|Q^{(0)}_{\rho'_\tau+1}(x) - \tilde{Q}_{\rho''_{\tau-1}}(x)| \leq |Q^{(0)}_{\rho'_\tau+1}(x) - \tilde{Q}_{\rho'_\tau}(x)| + |\tilde{Q}_{\rho'_\tau}(x) - \tilde{Q}_{\rho''_{\tau-1}}(x)|$$

$$\leq \frac{1}{2^{r_{\tau-1}}} + \frac{6}{2^{r_\tau}} + |F_{r_{\tau-1}}(x) - F_{r_\tau}(x)|$$

$$\left(x \in (\Omega_{r_{\tau-1}} \cap \Omega_{r_\tau} \cap \Omega_{r_\tau - 1}), \tau = 2, 3, \dots\right),$$

whence by (5.61)

$$|Q^{(0)}_{\rho'_\tau+1}(x) - Q^{(0)}_{\rho'_\tau}(x)| \leq \frac{1}{2^{r_{\tau-1}}} + \frac{6}{2^{r_\tau}} + |F_{r_{\tau-1}}(x) - F_{r_\tau}(x)|$$

$$\left(x \in (\Omega_{r_{\tau-1}} \cap \Omega_{r_\tau} \cap \Omega_{r_\tau - 1}), \tau = 2, 3, \dots\right).$$

Since the sequence (4.4) increases, it follows from the last inequality and (5.60), (5.61) that

$$|Q^{(0)}_t(x) - Q^{(0)}_{t-1}(x)| < \frac{7}{2^{r_{\tau-1}}} + |F_{r_{\tau-1}}(x) - F_{r_\tau}(x)|$$

$$\left(x \in (\Omega_{r_{\tau-1}} \cap \Omega_{r_\tau} \cap \Omega_{r_\tau - 1}), \rho''_{\tau-1} < t \leq \rho''_\tau, \tau = 2, 3, \dots\right). \tag{5.90}$$

Now define τ as a function of t by

$$\rho''_{\tau-1} < t \leq \rho''_\tau, \tag{5.91}$$

where $\tau \geq 1$. Then $\tau = \tau(t)$ is defined for $t > \rho''_0$ and

$$\lim_{t\to\infty} \tau = +\infty, \tag{5.92}$$

but since the sequence (4.4) of integers r_τ increases, we have by (5.92)

$$\lim_{t\to\infty} r_\tau = +\infty. \tag{5.93}$$

Now let e_0 denote the set of points x on $[-\pi, \pi]$ at which both equations (4.5) hold. Then

$$\text{mes } e_0 = 2\pi, \quad e_0 \subset [-\pi, \pi], \tag{5.94}$$

and, by (4.5) and (5.93),

$$\lim_{t\to\infty} \left[F_{r_{\tau-1}}(x) - F_{r_\tau}(x)\right] = 0 \quad (x \in e_0). \tag{5.95}$$

Put

$$D_t(x) = \frac{7}{2^{r_{\tau-1}}} + |F_{r_{\tau-1}}(x) - F_{r_\tau}(x)| \quad (x \in e_0, t > \rho''_1), \tag{5.96}$$

whence by (5.90), (5.93) and (5.95),

$$|Q^{(0)}_t(x) - Q^{(0)}_{t-1}(x)| < D_t(x) \quad \left(x \in (\Omega_{r_{\tau-1}} \cap \Omega_{r_\tau} \cap \Omega_{r_\tau - 1}),\ \ x \in e_0, t > \rho''_1\right), \tag{5.97}$$

$$\lim_{t\to\infty} D_t(x) = 0 \quad (x \in e_0). \tag{5.98}$$

We then put

$$W_t = (\Omega_{r_{\tau-1}} \cap \Omega_{r_\tau} \cap \Omega_{r_\tau - 1}), \quad W_t' = [-\pi, \pi] \setminus W_t \qquad (t > \rho_1''), \tag{5.99}$$

where τ is defined as a function of t by (5.91).

Hence we have, by (3.20) and (4.6),

$$\text{mes } W_t' < \frac{1}{2^{r_{\tau-1}}} + \frac{1}{2^{r_\tau}} + \frac{1}{2^{r_\tau - 1}} < \frac{4}{2^{r_{\tau-1}}}, \quad W_t' \in [-\pi, \pi] \quad (t > \rho_1''), \tag{5.100}$$

where W_t' is equal to $W_{\rho_\tau''}'$ for $\rho_{\tau-1}'' < t \leqslant \rho_\tau''$, and, since the sequence (4.4) increases,

$$\sum_{\tau=1}^{\infty} \frac{1}{2^{r_{\tau-1}}} < \infty. \tag{5.101}$$

Consequently if we put

$$W' = \varlimsup_{t\to\infty} W_t', \tag{5.102}$$

then by (5.100) and the remark after that condition we have

$$\text{mes } W' = 0, \quad W' \in [-\pi, \pi]. \tag{5.103}$$

Let x be any number satisfying

$$x \notin W', \quad x \in e_0 \subset [-\pi, \pi]. \tag{5.104}$$

Then by (5.102) there is an integer $t_0(x)$ such that

$$x \notin W_t', \quad x \in [-\pi, \pi] \quad (t > t_0(x)), \tag{5.105}$$

whence by (5.99)

$$x \in W_t \quad (t > t_0(x)), \tag{5.106}$$

and then it will follow from (5.97) and (5.99) that

$$|Q_t^{(0)}(x) - Q_{t-1}^{(0)}(x)| < D_t(x) \quad (t > t_1(x)), \tag{5.107}$$

if x satisfies (5.104), with

$$t_1(x) = \max[t_0(x), \rho_1'']. \tag{5.108}$$

Comparing (5.107) with (5.94), (5.98), (5.103) and (5.104), we see that

$$\lim_{t\to\infty} [Q_t^{(0)}(x) - Q_{t-1}^{(0)}(x)] = 0 \tag{5.109}$$

almost everywhere on $[-\pi, \pi]$.

If we replace $\tau - 1$ by τ in (5.85), we get

$$|\tilde{Q}_{\rho_\tau''}(x) - F_{r_\tau}(x)| < \frac{1}{2^{r_\tau}} \quad (x \in \Omega_{r_\tau}, \tau = 0, 1, 2, \ldots). \tag{5.110}$$

Here, by (3.20),

$$\text{mes } \Omega_{r_\tau} > 2\pi - \frac{1}{2^{r_\tau}}, \quad \Omega_{r_\tau} \subset [-\pi, \pi] \quad (\tau = 0, 1, 2, \ldots). \tag{5.111}$$

We also take

$$\Omega_\tau' = [-\pi, \pi] \setminus \Omega_{r_\tau} \quad (\tau = 0, 1, 2, \ldots), \tag{5.112}$$

$$\Omega^{(0)} = \varlimsup_{\tau\to\infty} \Omega_\tau'. \tag{5.113}$$

Then, by (5.101), (5.111) and (5.112),

$$\sum_{\tau=0}^{\infty} \text{mes}\, \Omega'_\tau < +\infty, \tag{5.114}$$

and it follows from (5.112) and (5.113) that

$$\text{mes}\, \Omega^{(0)} = 0, \quad \Omega^{(0)} \subset [-\pi, \pi]. \tag{5.115}$$

Now take an arbitrary x satisfying

$$x \in [-\pi, \pi] \setminus \Omega^{(0)}. \tag{5.116}$$

By (5.113) and (5.116) there is an integer

$$\tau_0 = \tau_0(x), \tag{5.117}$$

depending, in general, on x, such that

$$x \notin \Omega'_\tau, \quad x \in [-\pi, \pi] \quad (\tau \geqslant \tau_0), \tag{5.118}$$

whence by (5.112)

$$x \in \Omega_{r_\tau} \quad (\tau \geqslant \tau_0). \tag{5.119}$$

Consequently we find from (5.110) that

$$|\tilde{Q}_{\rho''_\tau}(x) - F_{r_\tau}(x)| < \frac{1}{2^{r_\tau}} \quad \left(x \in [-\pi, \pi] \setminus \Omega^{(0)}, \quad \tau \geqslant \tau_0\right). \tag{5.120}$$

Comparing the last inequality with (5.115), we see that

$$\lim_{\tau \to \infty} \left[\tilde{Q}_{\rho''_\tau}(x) - F_{r_\tau}(x)\right] = 0 \tag{5.121}$$

almost everywhere on $[-\pi, \pi]$. It follows from the first equation (4.5), the equation $\tilde{Q}_{\rho''_{\tau-1}}(x) = \tilde{Q}^{(0)}_{\rho''_{\tau-1}}(x)$ for $\tau = 2, 3, \ldots$ (see (5.61)), and (5.121), that

$$\lim_{\tau \to \infty} Q^{(0)}_{\rho''_\tau}(x) = F(x) \tag{5.122}$$

almost everywhere on $[-\pi, \pi]$. On the other hand, since (by (5.34)) the sequence of integers

$$\rho'_\tau \quad (\tau = 0, 1, 2, \ldots), \quad \rho''_\tau \quad (\tau = 0, 1, 2, \ldots) \tag{5.123}$$

increases, we have by (5.122)

$$\overline{\lim_{t \to \infty}}\; Q^{(0)}_t(x) \geqslant F(x) \tag{5.124}$$

almost everywhere on $[-\pi, \pi]$.

Put

$$\lambda(r, v) = \lambda_{r, \mu_r + v} \quad (r = 0, 1, 2, \ldots; v = 0, 1, 2, \ldots), \tag{5.125}$$

$$i_\tau = \Pi_{r_\tau - 1} + \lambda(r_\tau, u'_\tau - 1) - l(r_\tau, 0) \quad (\tau = 0, 1, 2, \ldots), \tag{5.126}$$

where $\lambda_{r,\mu}$ (see (3.8)) is defined by the second condition (3.12), and $l(r, v)$ and u'_τ are defined, respectively, by (3.35) and (5.64). Then

$$\lambda(r_\tau, u'_\tau - 1) = l(r_\tau, 0) + i_\tau - \Pi_{r_\tau - 1} \quad (\tau = 0, 1, 2, \ldots). \tag{5.127}$$

By (4.6), (3.13), (3.35), (5.125) and (5.65),

$$l(r_\tau, 0) \leqslant l(r_\tau, u'_\tau - 1) < \lambda(r_\tau, u'_\tau - 1) < l(r_\tau, u'_\tau) \leqslant l(r_\tau, 2^{r_\tau - 2})$$
$$(\tau = 0, 1, 2, \ldots), \tag{5.128}$$

whence by (3.31), (3.33), (3.35), (5.125) and (5.126),

$$\Pi_{r_\tau - 1} < i_\tau < \Pi_{r_\tau} \quad (\tau = 0, 1, 2, \dots), \tag{5.129}$$

and consequently we obtain, by (4.6), (3.36), (3.37) and (5.127),

$$r_\tau > 2, \quad \tilde{Q}_{i_\tau}(x) = Q_{r_\tau, \lambda(r_\tau, u'_\tau - 1)}(x) \quad (\tau = 0, 1, 2, \dots). \tag{5.130}$$

In addition, we obtain from the first relation (5.130), by (3.50), (4.22), (5.64) and (5.65),

$$t'_{r_\tau} = \Pi_{r_\tau - 1} + l(r_\tau, u'_\tau - 1) - l(r_\tau, 0) \quad (\tau = 0, 1, 2, \dots), \tag{5.131}$$

$$t''_{r_\tau} = \Pi_{r_\tau - 1} + l(r_\tau, u'_\tau) - l(r_\tau, 0) \quad (\tau = 0, 1, 2, \dots), \tag{5.132}$$

and consequently, by (5.33), (5.126) and (5.128),

$$\rho'_\tau = t'_{r_\tau} < i_\tau < t''_{r_\tau} = \rho''_\tau \quad (\tau = 0, 1, 2, \dots). \tag{5.133}$$

It then follows by (5.34), (5.35) and (5.38) that

$$Q^{(0)}_{i_\tau}(x) = \tilde{Q}_{i_\tau}(x) \quad (\tau = 1, 2, \dots). \tag{5.134}$$

Comparing the last equation with (5.130), we obtain

$$Q^{(0)}_{i_\tau}(x) = Q_{r_\tau, \lambda(r_\tau, u'_\tau - 1)}(x) \quad (\tau = 1, 2, \dots). \tag{5.135}$$

Since the sequence (5.123) increases, we have from (5.133)

$$\lim_{\tau \to \infty} i_\tau = \infty. \tag{5.136}$$

In addition, if we take account of the second inequality (3.23), (5.65) and (5.125), we obtain

$$|Q_{r_\tau, \lambda(r_\tau, u'_\tau - 1)}(x) - G_{r_\tau}(x)| < \frac{1}{2^{r_\tau}} \quad (x \in \Omega_{r_\tau}, \tau = 0, 1, 2, \dots), \tag{5.137}$$

whence by (5.135)

$$|Q^{(0)}_{i_\tau}(x) - G_{r_\tau}(x)| < \frac{1}{2^{r_\tau}} \quad (x \in \Omega_{r_\tau}, \tau = 0, 1, 2, \dots). \tag{5.138}$$

We now suppose that x satisfies (5.116). Then, as we have already proved, there is a number $\tau_0 = \tau_0(x)$ (see (5.117)) such that (5.119) is satisfied for the given value of x. In addition, since the sequence (4.4) increases, we have

$$\lim_{\tau \to \infty} r_\tau = \infty, \tag{5.139}$$

and then by (5.119) and (5.138) we obtain, for each x satisfying (5.116),

$$\lim_{\tau \to \infty} \left[Q^{(0)}_{i_\tau}(x) - G_{r_\tau}(x) \right] = 0. \tag{5.140}$$

Consequently, by (5.115), (5.140) is satisfied almost everywhere on $[-\pi, \pi]$.

Comparing the last equation with the second equation (4.5), we see that

$$\lim_{\tau \to \infty} Q^{(0)}_{i_\tau}(x) = G(x) \tag{5.141}$$

almost everywhere on $[-\pi, \pi]$, whence it follows, by (5.92) and (5.136), that

$$\underline{\lim}_{i \to \infty} Q^{(0)}_{i}(x) \leqslant G(x) \tag{5.142}$$

almost everywhere on $[-\pi, \pi]$.

Consider a function $\rho' = \rho'(t)$ which takes integral values and is defined by

$$\Pi_{\rho'-1} < \alpha(t) \leqslant \Pi_{\rho'}, \tag{5.143}$$

where $\alpha(t)$ is defined by (5.35). By (5.35) and (5.36), the function $\rho' = \rho'(t)$ is well defined and satisfies

$$\rho'(t+1) \geqslant \rho'(t) \tag{5.144}$$

for $t \geqslant t_0$, where t_0 is a sufficiently large integer. We may also require that

$$t_0 > \rho_0'', \tag{5.145}$$

where ρ_τ'' $(\tau = 0, 1, 2, \dots)$ is defined by (5.33). In addition, by (5.37),

$$\lim_{t\to\infty} \rho'(t) = \infty. \tag{5.146}$$

Comparing (3.13) and (3.35) with (5.143), we see that

$$l(\rho', 0) + \alpha(t) - \Pi_{\rho'-1} > l(\rho', 0) \geqslant \mu_{\rho'} \quad (t > t_0) \tag{5.147}$$

and consequently, by (3.24),

$$G_{\rho'}(x) - \frac{1}{2^{\rho'}} < Q_{\rho', l(\rho',0)+\alpha(t)-\Pi_{\rho'-1}}(x) < F_{\rho'}(x) + \frac{1}{2^{\rho'}} \quad (x \in \Omega_{\rho'}, t > t_0). \tag{5.148}$$

By (3.36), (3.37), (5.143) and (5.148), we obtain

$$G_{\rho'}(x) - \frac{1}{2^{\rho'}} < \tilde{Q}_{\alpha(t)}(x) < F_{\rho'}(x) + \frac{1}{2^{\rho'}} \quad (x \in \Omega_{\rho'}, t > t_0), \tag{5.149}$$

whence, by (5.38) and (5.145),

$$G_{\rho'}(x) - \frac{1}{2^{\rho'}} < Q_t^{(0)}(x) < F_{\rho'}(x) + \frac{1}{2^{\rho'}} \quad (x \in \Omega_{\rho'}, t > t_0). \tag{5.150}$$

It also follows from (5.34) and (5.35) that

$$\Pi_{r_\tau - 1} < \alpha(t) \leqslant \Pi_{r_\tau} \quad (\rho_\tau' < t \leqslant \rho_\tau'', \tau = 1, 2, \dots), \tag{5.151}$$

$$\Pi_{r_{\tau-1}-1} < \alpha(t) \leqslant \Pi_{r_{\tau-1}} \quad (\rho_{\tau-1}'' < t \leqslant \rho_\tau', \tau = 1, 2, \dots), \tag{5.152}$$

and since the integral-valued function $\rho' = \rho'(t)$ is defined by (5.143), we have

$$\rho' = \rho'(t) = r_\tau \quad (\rho_\tau' < t \leqslant \rho_\tau'', \tau = 1, 2, \dots, t > t_0), \tag{5.153}$$

$$\rho' = \rho'(t) = r_{\tau-1} \quad (\rho_{\tau-1}'' < t \leqslant \rho_\tau', \tau = 1, 2, \dots, t > t_0). \tag{5.154}$$

If we now consider $\tau = \tau(t)$ as a function of t defined by (5.91), equation (5.92) will be satisfied, and the function will be defined for $t > \rho_0''$. Then by (4.5) and (5.153), (5.154),

$$\lim_{t\to\infty} F_{\rho'}(x) = F(x), \quad \lim_{t\to\infty} G_{\rho'}(x) = G(x) \tag{5.155}$$

almost everywhere on $[-\pi, \pi]$.

Put

$$W_t'' = [-\pi, \pi] \setminus \Omega_{\rho'} \quad (t > t_0 > \rho_0''). \tag{5.156}$$

Then, by (3.20) and (5.153), (5.154),

$$\operatorname{mes} W_t'' < \frac{1}{2^{r_\tau}} \quad (\rho_\tau' < t \leqslant \rho_\tau'', \tau = 1, 2, \dots, t > t_0), \tag{5.157}$$

$$\operatorname{mes} W_t'' < \frac{1}{2^{r_{\tau-1}}} \quad (\rho_{\tau-1}'' < t \leqslant \rho_\tau', \tau = 1, 2, \dots, t > t_0), \tag{5.158}$$

where the sets W_t'' coincide for $\rho_\tau' < t \leqslant \rho_\tau''$ and equal W_{r_τ}'', and likewise coincide for $\rho_{\tau-1}'' < t \leqslant \rho_\tau'$ and equal $W_{r_{\tau-1}}''$. We then put

$$W'' = \overline{\lim_{t\to\infty}}\ W_t''. \tag{5.159}$$

Then, by (5.101), (5.156)–(5.158), and the remark after (5.158),

$$\operatorname{mes} W'' = 0, \quad W'' \subset [-\pi, \pi]. \tag{5.160}$$

Now let x be an arbitrary point satisfying

$$x \in [-\pi, \pi] \setminus W''. \tag{5.161}$$

Then by (5.159) there is an integer $t_1 = t_1(x)$, in general depending on x, such that

$$x \in [-\pi, \pi] \setminus W_t'' \quad (t > t_1). \tag{5.162}$$

Here we may suppose that

$$t_1 > t_0. \tag{5.163}$$

Hence it follows, by (5.145) and (5.156), that

$$x \in \Omega_{\rho'} \quad (t > t_1), \tag{5.164}$$

and then by (5.150)

$$G_{\rho'}(x) - \frac{1}{2^{\rho'}} < Q_t^{(0)}(x) < F_{\rho'}(x) + \frac{1}{2^{\rho'}} \quad (t > t_1 = t_1(x)) \tag{5.165}$$

for an arbitrary x satisfying (5.161).

On the other hand, since $\rho' = \rho'(t)$ satisfies (5.153) and (5.154), then because the sequence (4.4) increases we have, by (4.5) and (5.146),

$$\lim_{t\to\infty}\left[F_{\rho'}(x) + \frac{1}{2^{\rho'}}\right] = F(x), \quad \lim_{t\to\infty}\left[G_{\rho'}(x) - \frac{1}{2^{\rho'}}\right] = G(x), \tag{5.166}$$

and then it follows from (5.160), (5.161) and (5.165) that

$$G(x) \leqslant \underline{\lim_{t\to\infty}}\ Q_t^{(0)}(x) \leqslant \overline{\lim_{t\to\infty}}\ Q_t^{(0)}(x) \leqslant F(x) \tag{5.167}$$

almost everywhere on $[-\pi, \pi]$. Comparing (5.167) with (5.124) and (5.142), we see that

$$\overline{\lim_{t\to\infty}}\ Q_t^{(0)}(x) = F(x), \quad \underline{\lim_{t\to\infty}}\ Q_t^{(0)}(x) = G(x) \tag{5.168}$$

almost everywhere on $[-\pi, \pi]$.

For arbitrary $t \geqslant 2$ we have defined the sets (3.84) and the polynomials (3.85) with properties 1)–6) (see (3.86)–(3.94) and the remark at the end of §3). In particular, we have verified (3.94), whose left-hand side, as well as the functions $s_{\theta,k}^{(t)}(x)$, are obtained from $B_m^{(t)}(x, T)$ and $s_k^{(t)}(x)$ (see (3.88) and (3.89)) by replacing a_j and b_j $(M_{t-1} < j < M_t')$ by $\theta_j a_j$ and $\theta_j b_j$, where θ_j $(M_{t-1} < j < M_t')$ are arbitrary real numbers.

Since (1.18) is a subseries of (1.5) (see the statement at the end of §4), we have

$$a_j' = \theta_j a_j, \quad b_j' = \theta_j b_j \quad (j = 1, 2, \ldots), \tag{5.169}$$

where each θ_j is 0 or 1. Then the functions $\tilde{s}_k^{(t)}(x)$ and $\tilde{B}_m^{(t)}(x, T)$ defined by (5.42) and (5.43) are obtained from the functions $s_{\theta,k}^{(t)}(x)$ and $B_{\theta,m}^{(t)}(x, T)$ by supposing that the numbers θ_j satisfy (5.169). Consequently, by (3.94),

$$\tilde{B}_m^{(t)}(x, T) = 0 \quad (x \in [-\pi, \pi], 0 \leqslant m \leqslant M_{t-1}, t \geqslant 2). \tag{5.170}$$

By (3.121) we have $M_t \leqslant M_{t'-1}$ if $t' > t$, and then it follows from (5.170) that

$$\tilde{B}_m^{(t')}(x, T) = 0 \quad (x \in [-\pi, \pi], \quad 0 \leqslant m \leqslant M_t, t' > t, t = 1, 2, \dots). \tag{5.171}$$

Comparing (5.26) and (5.27) with (5.41) and (5.42), we obtain

$$s'_{t-1,k}(x) + \tilde{s}_k^{(t)}(x) = \begin{cases} \sum_{j=1}^{k} (a'_j \cos jx + b'_j \sin jx) & (0 \leqslant k \leqslant M_{t-1}), \\ H'_{t-1}(x) + \sum_{j=M_{t-1}+1}^{k} (a'_j \cos jx + b'_j \sin jx) & \\ \qquad (M_{t-1} < k \leqslant M_t), & \\ H'_{t-1}(x) + \tilde{H}'_t(x) \quad (M_t < k) & \end{cases}$$

$$(t = 3, 4, \dots). \tag{5.172}$$

In addition, by (5.26),

$$H'_{t-1}(x) + \sum_{j=M_{t-1}+1}^{k} (a'_j \cos jx + b'_j \sin jx) = \sum_{j=1}^{k} (a'_j \cos jx + b'_j \sin jx)$$

$$(M_{t-1} < k \leqslant M_t, \quad t = 3, 4, \dots), \tag{5.173}$$

and by (5.26), (5.41) and (5.7),

$$H'_{t-1}(x) + \tilde{H}'_t(x) = \sum_{j=1}^{M_t} (a'_j \cos jx + b'_j \sin jx) = H'_t(x) \quad (t = 3, 4, \dots). \tag{5.174}$$

It follows from the last three equations and (5.8) that

$$s'_{t-1,k}(x) + \tilde{s}_k^{(t)}(x) = s'_{t,k}(x) \quad (t = 3, 4, \dots, \ k = 0, 1, 2, \dots). \tag{5.175}$$

We now establish the equation

$$s'_{t-1,k}(x) + \sum_{t'=t}^{t''} \tilde{s}_k^{(t')}(x) = s'_{t'',k}(x) \quad (t'' \geqslant t, t \geqslant 3, k = 0, 1, 2, \dots). \tag{5.176}$$

If $t'' = t$, (5.176) is valid for arbitrary $t \geqslant 3$, since it coincides with (5.175). Suppose now that (5.176) is valid for some $t'' \geqslant t$. We shall show that it remains valid if we replace t'' by $t'' + 1$.

By (5.175) and (5.176),

$$\begin{aligned} s'_{t-1,k}(x) + \sum_{t'=1}^{t''+1} \tilde{s}_k^{(t')}(x) &= s'_{t-1,k}(x) + \sum_{t'=1}^{t''} \tilde{s}_k^{(t')}(x) + \tilde{s}_k^{(t''+1)}(x) \\ &= s'_{t'',k}(x) + \tilde{s}_k^{(t''+1)}(x) = s'_{t''+1,k}(x) \end{aligned}$$

$$(t \geqslant 3, k = 0, 1, 2, \dots).$$

Consequently (5.176) is valid for $t \geqslant 3$ and $k = 0, 1, 2, \dots$, if we replace t'' by $t'' + 1$; and so, by induction, (5.176) is valid if $t'' \geqslant t$, $t \geqslant 3$ and $k = 0, 1, 2, \dots$.

Since (see the statement of Theorem 1 in §1) the method T was assumed row-finite, we can find for each $m = 0, 1, 2, \dots$ an integer t_m such that

$$c_{mk} = 0 \quad (k \geqslant t_m, m = 0, 1, 2, \dots), \tag{5.177}$$

whence, by (1.20), (5.9) and (5.43),

$$T_m^{(1)}(x) = \sum_{k=1}^{t_m} c_{mk} s'_k(x) \qquad (m = 0, 1, 2, \dots), \tag{5.178}$$

$$B'_{t,m}(x, T) = \sum_{k=0}^{t_m} c_{mk} s'_{t,k}(x) \qquad (m = 0, 1, 2, \dots, t \geqslant 2), \tag{5.179}$$

$$\tilde{B}_m^{(t)}(x, T) = \sum_{k=0}^{t_m} c_{mk} \tilde{s}_m^{(t)}(x) \qquad (m = 0, 1, 2, \dots; t \geqslant 2). \tag{5.180}$$

Comparing (5.176) with (5.179) and (5.180), we obtain

$$B'_{t-1,m}(x, T) + \sum_{t'=t}^{t''} \tilde{B}_m^{(t')}(x, T) = B'_{t'',m}(x, T)$$
$$(t'' \geqslant t;\ t = 3, 4, \dots;\ m = 0, 1, 2, \dots). \tag{5.181}$$

In particular, for $t'' = t$ we get

$$B'_{t-1,m}(x, T) + \tilde{B}_m^{(t)}(x, T) = B'_{t,m}(x, T) \quad (t = 3, 4, \dots;\ m = 0, 1, 2, \dots). \tag{5.182}$$

Put

$$\kappa_t = t + 1 + \max_{0 \leqslant m \leqslant M_t} t_m \quad (t = 1, 2, \dots), \tag{5.183}$$

whence

$$\kappa_t > t \quad (t = 1, 2, \dots), \tag{5.184}$$

and consequently

$$\kappa_t \geqslant 2 \quad (t = 1, 2, \dots). \tag{5.185}$$

In addition, if we take account of (5.171) and (5.184), we obtain

$$\sum_{t'=t}^{\kappa_t} \tilde{B}_m^{(t')}(x, T) = \tilde{B}_m^{(t)}(x, T) \quad (0 \leqslant m \leqslant M_t, t = 1, 2, \dots), \tag{5.186}$$

and then, by (5.181), (5.182) and (5.184),

$$B'_{\kappa_t,m}(x, T) = B'_{t-1,m}(x, T) + \tilde{B}_m^{(t)}(x, T) = B'_{t,m}(x, T) \tag{5.187}$$
$$(0 \leqslant m \leqslant M_t, t = 3, 4, \dots).$$

It follows from the preceding inequality and from (5.179) and (5.185) that

$$B'_{t,m}(x, T) = \sum_{k=1}^{t_m} c_{mk} s'_{\kappa_t,k}(x) \quad (0 \leqslant m \leqslant M_t;\ t \geqslant 3). \tag{5.188}$$

Taking account of (5.8) and (5.185), we obtain

$$s'_{\kappa_t,k}(x) = \sum_{j=1}^{k} (a'_j \cos jx + b'_j \sin jx) \qquad (0 \leqslant k \leqslant M_{\kappa_t}, t = 1, 2, \dots), \tag{5.189}$$

whence, by (1.21),

$$s'_{\kappa_t,k}(x) = s'_k(x) \qquad (0 \leqslant k \leqslant M_{\kappa_t}, t = 1, 2, \dots). \tag{5.190}$$

In addition, since the second inequality (3.122) is valid for $t = 2, 3, \dots$ (see the remark at the end of §3), we find from (5.183) and (5.185) that

$$t_m < \kappa_t \leqslant M_{\kappa_t} \qquad (0 \leqslant m \leqslant M_t, t = 1, 2, \dots), \tag{5.191}$$

and then, if for some $t = 1, 2, \dots$ the integers k and m satisfy

$$0 \leqslant k \leqslant t_m, \quad 0 \leqslant m \leqslant M_t, \tag{5.192}$$

we have

$$0 \leqslant k \leqslant M_{\kappa_t} \quad (t = 1, 2, \dots). \tag{5.193}$$

Hence it follows, by (5.178) and (5.190), that

$$T_m^{(1)}(x) = \sum_{k=1}^{t_m} c_{mk} s'_{\kappa_t,k}(x) \qquad (0 \leqslant m \leqslant M_t, t = 1, 2, \dots).$$

Comparing the preceding inequality with (5.188) and (5.187), we obtain

$$T_m^{(1)}(x) = B'_{t,m}(x, T) \quad (0 \leqslant m \leqslant M_t, t \geqslant 3), \tag{5.194}$$

$$T_m^{(1)}(x) = B'_{t-1,m}(x, T) + \tilde{B}_m^{(t)}(x, T) \quad (0 \leqslant m \leqslant M_t, t = 3, 4, \dots). \tag{5.195}$$

Since, by (5.35),

$$\alpha(t) = \rho''_\tau = \alpha(\rho''_\tau) \quad (\rho''_\tau < t \leqslant \rho'_{\tau+1}, \tau = 1, 2, \dots),$$
$$\alpha(t) = t \quad (\rho'_\tau < t \leqslant \rho''_\tau, \tau = 1, 2, \dots), \tag{5.196}$$

it follows from (5.51) that

$$\tilde{B}_m^{(\alpha(t))}(x, T) = B_m^{(\alpha(t))}(x, T) \quad (\rho'_2 < t, m = 0, 1, 2, \dots). \tag{5.197}$$

In addition, if we take account of (5.34) we have $\rho''_\tau > \rho'_\tau \geqslant 3$ $(\tau = 2, 3, \dots)$, whence by (5.196)

$$\alpha(t) \geqslant 3 \quad (\rho'_2 < t). \tag{5.198}$$

From (4.6), (4.22), (5.33) and (5.64) we obtain

$$r_\tau > 2, \quad \rho'_\tau = t_{r_\tau,0,u'_\tau}, \quad \rho''_\tau = t_{r_\tau,1,u'_\tau} \quad (\tau = 0, 1, 2, \dots). \tag{5.199}$$

Then if

$$\rho'_\tau < t \leqslant \rho''_\tau \quad (\tau = 1, 2, \dots), \tag{5.200}$$

we have, after taking account of (5.34) and (5.65),

$$r_\tau = r(t), \quad u'_\tau = h(t) \quad (\rho'_\tau < t \leqslant \rho''_\tau, \tau = 2, 3, \dots), \tag{5.201}$$

where $r = r(t)$ and $h = h(t)$ are the functions of t defined by (3.69)–(3.71).

Since

$$\rho'_\tau \geqslant 2 \quad (\tau = 1, 2, \dots) \tag{5.202}$$

by (5.34), it follows from (5.16), (5.33), (5.64) and (5.201) that

$$B'_{t-1,m}(x, T) = B_{t-1,m}(x, h(t), T) \quad (\rho'_\tau < t \leqslant \rho''_\tau; \tau = 2, 3, \dots; m = 0, 1, 2, \dots), \tag{5.203}$$

and since (3.81) is valid for $t \geqslant 2$, and $h = h(t)$ in that inequality, by (5.35), (5.202), and (5.203) we obtain from (3.81)

$$\beta_{\alpha(t)}(x) = \tilde{Q}_{\alpha(t)}(x) - B'_{\alpha(t)-1, M_{\alpha(t)-1}}(x, T) \quad (\rho'_\tau < t \leqslant \rho''_\tau, \tau = 2, 3, \ldots). \quad (5.204)$$

Taking account of the first equation (5.196), we see that (5.204) is also valid for t satisfying

$$\rho''_\tau < t \leqslant \rho'_{\tau+1} \quad (\tau = 2, 3, \ldots), \quad (5.205)$$

and then

$$\beta_{\alpha(t)}(x) = \tilde{Q}_{\alpha(t)}(x) - B'_{\alpha(t)-1, M_{\alpha(t)-1}}(x, T) \quad (\rho'_2 < t) \quad (5.206)$$

and consequently, by (5.197),

$$\beta_{\alpha(t)}(x) - B^{(\alpha(t))}_{M_{\alpha(t)}}(x, T) = \tilde{Q}_{\alpha(t)}(x) - B'_{\alpha(t)-1, M_{\alpha(t)-1}}(x, T) - \tilde{B}^{(\alpha(t))}_{M_{\alpha(t)}}(x, T) \quad (\rho'_2 < t). \quad (5.207)$$

Comparing the last equation with (5.195) and (5.198), we obtain

$$\begin{aligned} \beta_{\alpha(t)}(x) - B^{(\alpha(t))}_{M_{\alpha(t)}}(x, T) &= \tilde{Q}_{\alpha(t)}(x) - B'_{\alpha(t)-1, M_{\alpha(t)}}(x, T) - \tilde{B}^{(\alpha(t))}_{M_{\alpha(t)}}(x, T) \\ &\quad + B'_{\alpha(t)-1, M_{\alpha(t)}}(x, T) - B'_{\alpha(t)-1, M_{\alpha(t)-1}}(x, T) \\ &= \tilde{Q}_{\alpha(t)}(x) - T^{(1)}_{\alpha(t)}(x) + B'_{\alpha(t)-1, M_{\alpha(t)}}(x, T) - B'_{\alpha(t)-1, M_{\alpha(t)-1}}(x, T) \quad (\rho'_2 < t). \end{aligned} \quad (5.208)$$

From (3.92) and the first inequality (3.119), we have

$$|B^{(t)}_m(x, T) - \beta_t(x)| < \frac{1}{2^t} \quad (x \in [-\pi, \pi] \setminus E_t, \quad m \geqslant M_t, t \geqslant 2), \quad (5.209)$$

and consequently, by (5.198),

$$|B^{(\alpha(t))}_{M_{\alpha(t)}}(x, T) - \beta_{\alpha(t)}(x)| < \frac{1}{2^{\alpha(t)}} \quad (x \in [-\pi, \pi] \setminus E_{\alpha(t)}, \quad \rho'_2 < t). \quad (5.210)$$

We have already seen that the numbers (4.27) are special instances of the numbers $\tilde{a}_0$, $\tilde{a}_j$ and $\tilde{b}_j$ $(j = 1, 2, \ldots)$ which satisfy (3.128) for arbitrary $t = 2, 3, \ldots$ (see the remark before (5.6)). Then all the conditions satisfied by the numbers $\tilde{a}_j$ and $\tilde{b}_j$ are also satisfied for the numbers (4.27). In particular, (3.120) is satisfied if $\tilde{a}_j$ and $\tilde{b}_j$ are replaced by a'_j and b'_j. Therefore if we compare (3.101), (3.114) and (5.9) with (3.120), we obtain

$$|B'_{t,m'}(x, T) - B'_{t,m''}(x, T)| < \frac{1}{2^t} \qquad (m', m'' \geqslant M_t; t = 2, 3, \ldots; -\pi \leqslant x \leqslant \pi). \quad (5.211)$$

Since $\alpha(t) \geqslant 3$ if $t > \rho'_2$ (see (5.198)), we have $\alpha(t) - 1 \geqslant 2$ for the same values of t, whence it follows by (3.121) and (5.211) that

$$|B'_{\alpha(t)-1, M_{\alpha(t)}}(x, T) - B'_{\alpha(t)-1, M_{\alpha(t)-1}}(x, T)| < \frac{1}{2^{\alpha(t)-1}} \quad (\rho'_2 < t). \quad (5.212)$$

Taking account of (5.208) and (5.210), we find from (5.212) that

$$|\tilde{Q}_{\alpha(t)}(x) - T^{(1)}_{\alpha(t)}(x)| < \frac{3}{2^{\alpha(t)}} \quad (x \in [-\pi, \pi] \setminus E_{\alpha(t)}, \rho'_2 < t). \quad (5.213)$$

Comparing the preceding inequality with (5.194), we see that

$$|\tilde{Q}_{\alpha(t)}(x) - B'_{\alpha(t), M_{\alpha(t)}}(x, T)| < \frac{3}{2^{\alpha(t)}} \quad (x \in [-\pi, \pi] \setminus E_{\alpha(t)}, \rho'_2 < t), \tag{5.214}$$

whence

$$|\tilde{Q}_{\alpha(t-1)}(x) - B'_{\alpha(t-1), M_{\alpha(t-1)}}(x, T)| < \frac{3}{2^{\alpha(t-1)}} \quad (x \in [-\pi, \pi] \setminus E_{\alpha(t-1)}, \rho'_2 + 1 < t). \tag{5.215}$$

If

$$\rho'_\tau + 1 < t \leqslant \rho''_\tau \quad (\tau = 2, 3, \dots), \tag{5.216}$$

then

$$\rho'_\tau < t - 1 < \rho''_\tau \quad (\tau = 2, 3, \dots), \tag{5.217}$$

and consequently, by (5.35),

$$\alpha(t) = t, \quad \alpha(t-1) = t - 1 = \alpha(t) - 1 \quad (\rho'_\tau + 1 < t \leqslant \rho''_\tau, \tau = 2, 3, \dots), \tag{5.218}$$

and then by (5.215)

$$|\tilde{Q}_{\alpha(t)-1}(x) - B'_{\alpha(t)-1, M_{\alpha(t)-1}}(x, T)| < \frac{6}{2^{\alpha(t)}}$$
$$(x \in [-\pi, \pi] \setminus E_{\alpha(t)-1}, \rho'_\tau + 1 < t \leqslant \rho''_\tau, \tau = 2, 3, \dots). \tag{5.219}$$

From (5.206) and (5.219) we get

$$|\beta_{\alpha(t)}(x)| \leqslant |\tilde{Q}_{\alpha(t)}(x) - \tilde{Q}_{\alpha(t)-1}(x)| + \frac{6}{2^{\alpha(t)}}$$
$$(x \in [-\pi, \pi] \setminus E_{\alpha(t)-1}; \rho'_\tau + 1 < t \leqslant \rho''_\tau, \tau = 2, 3, \dots), \tag{5.220}$$

and it follows from (3.47), (4.6) and (5.34) that

$$|\tilde{Q}_t(x) - \tilde{Q}_{t-1}(x)| < \frac{1}{2^{r_\tau - 2}} \quad (x \in \Omega_{r_\tau}, \rho'_\tau < t \leqslant \rho''_\tau, \tau = 1, 2, \dots), \tag{5.221}$$

whence, by (5.35) and (5.34),

$$|\tilde{Q}_{\alpha(t)}(x) - \tilde{Q}_{\alpha(t)-1}(x)| < \frac{4}{2^{r_\tau}} \quad (x \in \Omega_{r_\tau}; \rho'_\tau < t \leqslant \rho''_\tau; \tau = 1, 2, \dots). \tag{5.222}$$

Hence, by (5.220) and (5.34), we obtain

$$\alpha(t) = t, \quad |\beta_{\alpha(t)}(x)| < \frac{4}{2^{r_\tau}} + \frac{6}{2^{\alpha(t)}} < \frac{4}{2^{r_\tau}} + \frac{6}{2^{\rho''_\tau - 1}}$$
$$(x \in [-\pi, \pi] \setminus E_{\alpha(t)-1}, x \in \Omega_{r_\tau}, \rho'_\tau + 1 < t \leqslant \rho''_\tau, \tau = 2, 3, \dots). \tag{5.223}$$

Now suppose that

$$\rho''_{\tau-1} < t \leqslant \rho'_\tau \quad (\tau = 3, 4, \dots). \tag{5.224}$$

By (5.223) and (5.35), we have

$$\alpha(t) = \rho''_{\tau-1}, \quad |\beta_{\alpha(t)}(x)| < \frac{4}{2^{r_{\tau-1}}} + \frac{6}{2^{\alpha(t)}}$$
$$(x \in [-\pi, \pi] \setminus E_{\alpha(t)-1}, x \in \Omega_{r_{\tau-1}}, \tau = 3, 4, \dots), \tag{5.225}$$

whence

$$|\beta_{\alpha(t)}| < \frac{4}{2^{r_{\tau-1}}} + \frac{6}{2^{\rho''_{\tau-1}}}$$

$$\left(x \in [-\pi, \pi] \setminus E_{\alpha(t)-1}, x \in \Omega_{r_{\tau-1}}, \rho''_{\tau-1} < t \leqslant \rho'_\tau, \tau = 3, 4, \dots\right). \tag{5.226}$$

Suppose finally that

$$t = \rho'_\tau + 1 \quad (\tau = 3, 4, \dots), \tag{5.227}$$

whence

$$t - 1 = \rho'_\tau \quad (\tau = 3, 4, \dots). \tag{5.228}$$

Then, by (5.35) and (5.227), we obtain

$$\alpha(t) = \rho'_\tau + 1, \quad \alpha(t) - 1 = \rho'_\tau \quad (\tau = 3, 4, \dots), \tag{5.229}$$

$$\alpha(t-1) = \rho''_{\tau-1} \qquad (\tau = 3, 4, \dots), \tag{5.230}$$

and then, by (5.215), (5.227) and the inequalities $\rho'_\tau > \rho''_{\tau-1}$ for $\tau = 2, 3, \dots$ (see (5.34)),

$$|\tilde{Q}_{\rho''_{\tau-1}}(x) - B'_{\rho''_{\tau-1}, M_{\rho''_{\tau-1}}}(x, T)| < \frac{3}{2^{\rho''_{\tau-1}}} \quad \left(x \in [-\pi, \pi] \setminus E_{\rho''_{\tau-1}}, \tau = 3, 4, \dots\right). \tag{5.231}$$

In addition, since it follows from (5.34) that

$$\rho''_{\tau-1} + 1 \geqslant 3, \quad \rho''_{\tau-1} < \rho'_\tau \quad (\tau = 1, 2, \dots), \tag{5.232}$$

we have by (5.181)

$$B'_{\rho''_{\tau-1}, M_{\rho'_\tau}}(x, T) + \sum_{t' = \rho''_\tau + 1}^{\rho'_\tau} \tilde{B}^{(t')}_{M_{\rho'_\tau}}(x, T) = B'_{\rho'_\tau, M_{\rho'_\tau}}(x, T) \quad (\tau = 1, 2, \dots), \tag{5.233}$$

whence, taking account of (5.50), we obtain

$$B'_{\rho''_{\tau-1}, M_{\rho'_\tau}}(x, T) = B'_{\rho'_\tau, M_{\rho'_\tau}}(x, T) \quad (\tau = 2, 3, \dots). \tag{5.234}$$

Comparing (5.211) with (5.232) and (5.234), we see that

$$|B_{\rho''_{\tau-1}, M_{\rho''_{\tau-1}}}(x, T) - B'_{\rho'_\tau, M_{\rho'_\tau}}(x, T)| < \frac{1}{2^{\rho''_{\tau-1}}} \quad (\tau = 2, 3, \dots; -\pi \leqslant x \leqslant \pi), \tag{5.235}$$

and then, by (5.231) and (5.232),

$$|\tilde{Q}_{\rho''_{\tau-1}}(x) - B'_{\rho'_\tau, M_{\rho'_\tau}}(x, T)| < \frac{4}{2^{\rho''_{\tau-1}}} \quad \left(x \in [-\pi, \pi] \setminus E_{\rho''_{\tau-1}}, \tau = 3, 4, \dots\right). \tag{5.236}$$

It follows from (5.65), (5.86) and (5.88) that

$$|\tilde{Q}_{\rho'_\tau}(x) - F_{r_\tau}(x)| < \frac{2}{2^{r_\tau}} \quad \left(x \in (\Omega_{r_\tau} \cap \Omega_{r_\tau - 1}), \tau = 1, 2, \dots\right). \tag{5.237}$$

Hence, taking account of (5.85) and (5.236), we have

$$|\tilde{Q}_{\rho'_\tau}(x) - B'_{\rho'_\tau, M_{\rho'_\tau}}(x, T)| < \frac{1}{2^{r_{\tau-1}}} + \frac{2}{2^{r_\tau}} + \frac{4}{2^{\rho''_{\tau-1}}} + \zeta_\tau(x)$$

$$\left(x \in [-\pi, \pi] \setminus E_{\rho''_{\tau-1}}, x \in e_0, x \in (\Omega_{r_\tau - 1} \cap \Omega_{r_\tau} \cap \Omega_{r_{\tau-1}}), \tau = 3, 4, \dots\right), \tag{5.238}$$

where e_0 is the set of points of $[-\pi, \pi]$ at which (4.5) is satisfied (see the definition preceding (5.94)), and

$$\zeta_\tau(x) = |F_{r_{\tau-1}}(x) - F_{r_\tau}(x)| \quad (x \in [-\pi, \pi], x \in e_0, \tau = 2, 3, \dots). \quad (5.239)$$

Now since (4.5) is satisfied for $x \in e_0$, we have

$$\lim_{\tau\to\infty} \zeta_\tau(x) = 0 \quad (x \in e_0). \quad (5.240)$$

We assumed that (5.227) is satisfied at t, and then, as we have seen, (5.229) is satisfied. Hence, by (4.6) and (5.238),

$$|\tilde{Q}_{\alpha(t)-1}(x) - B'_{\alpha(t)-1, M_{\alpha(t)-1}}(x, T)| < \frac{3}{2^{r_{\tau-1}}} + \frac{4}{2^{\rho''_{\tau-1}}} + \zeta_\tau(x)$$

$$\left(t = \rho'_\tau + 1;\ x \in [-\pi, \pi] \setminus E_{\rho''_{\tau-1}}, x \in e_0, x \in (\Omega_{r_{\tau-1}} \cap \Omega_{r_\tau} \cap \Omega_{r_\tau-1}), \tau = 3, 4, \dots\right). \quad (5.241)$$

In addition, by (5.204),

$$|\beta_{\alpha(t)}(x)| = |\tilde{Q}_{\alpha(t)}(x) - B'_{\alpha(t)-1, M_{\alpha(t)-1}}(x, T)|$$
$$\leqslant |\tilde{Q}_{\alpha(t)}(x) - \tilde{Q}_{\alpha(t)-1}(x)| + |\tilde{Q}_{\alpha(t)-1}(x) - B'_{\alpha(t)-1, M_{\alpha(t)-1}}(x)|$$
$$(\rho'_\tau < t \leqslant \rho''_\tau, \tau = 2, 3, \dots),$$

whence, taking account of (5.34), (5.222) and (5.241), we have

$$|\beta_{\alpha(t)}(x)| < \frac{4}{2^{r_\tau}} + \frac{3}{2^{r_{\tau-1}}} + \frac{4}{2^{\rho''_{\tau-1}}} + \zeta_\tau(x)$$

$$\left(t = \rho'_\tau + 1, \quad x \in [-\pi, \pi] \setminus E_{\rho''_{\tau-1}}, x \in e_0, x \in (\Omega_{r_{\tau-1}} \cap \Omega_{r_\tau} \cap \Omega_{r_\tau-1}), \tau = 3, 4, \dots\right). \quad (5.242)$$

From now on we shall consider τ to be the function of t defined by (5.91) for $t > \rho''_0$. Then, as we have seen, (5.92) and (5.93) are satisfied, and since the sequence (5.123) increases, we have

$$\lim_{\tau\to\infty} \rho'_\tau = \infty, \quad \lim_{\tau\to\infty} \rho''_\tau = \infty. \quad (5.243)$$

Consequently, taking account of (5.136),

$$\lim_{t\to\infty} \rho'_\tau = \infty, \quad \lim_{t\to\infty} \rho''_{\tau-1} = \infty, \quad \lim_{t\to\infty} i_\tau = \infty. \quad (5.244)$$

Now define $\omega(t)$ as a function of t by

$$\omega(t) = \frac{4}{2^{r_\tau}} + \frac{4}{2^{r_{\tau-1}}} + \frac{6}{2^{\rho''_{\tau-1}}} + \frac{6}{2^{\alpha(t)}}, \quad (5.245)$$

where $\tau = \tau(t)$ is the function of t defined above. Then since $\alpha(t)$ is defined for $t > \rho'_0$ (see the remark before (5.35)), the function $\omega(t)$ is defined for $t > \rho''_0$, and, by (5.37), (5.93), (5.244) and (5.245),

$$\lim_{t\to\infty} \omega(t) = 0. \quad (5.246)$$

Supposing as before that $\tau = \tau(t)$ is defined for $t > \rho''_0$, as a function of t, by (5.91), and taking into account that

$$\rho''_{\tau-1} < \rho''_\tau \quad (\tau = 1, 2, \dots) \quad (5.247)$$

(see (5.34)), we introduce the notation

$$E_t' = E_{\alpha(t)-1} \cup E_{\alpha(t)} \cup E_{\rho''_{\tau-1}} \cup W_t' \quad (t > \rho_2''), \tag{5.248}$$

where W_t' is defined by (5.99). Then we have, by (5.223), (5.226), (5.242) and (5.245), and the inequalities $r_{\tau-1} < r_\tau$, $\tau = 1, 2, \ldots$ (see (4.6)),

$$|\beta_{\alpha(t)}(x)| < \omega(t) + \zeta_\tau(x) \quad (x \in [-\pi, \pi] \setminus E_t', x \in e_0, t > \rho_2''). \tag{5.249}$$

Now put

$$E' = \overline{\lim_{t\to\infty}}\, E_t', \quad E_0' = \overline{\lim_{t\to\infty}}\, E_t, \tag{5.250}$$

$$E_1' = \overline{\lim_{t\to\infty}}\, E_{\alpha(t)-1}, \quad E_2' = \overline{\lim_{t\to\infty}}\, E_{\alpha(t)}, \quad E_3' = \overline{\lim_{t\to\infty}}\, E_{\rho''_{\tau-1}}, \tag{5.251}$$

whence it follows by (5.102) and (5.248) that

$$E' = E_1' \cup E_2' \cup E_3' \cup W'. \tag{5.252}$$

Since, by (5.247), the sequence of integers $\rho''_{\tau-1}$ $(\tau = 1, 2, \ldots)$ increases, the second equation (5.250) and third equation (5.251) imply that

$$E_3' \subset E_0'. \tag{5.253}$$

In addition, we see from (5.36) and (5.37) that the sequences of integers

$$\alpha(t) \quad (t > \rho_0''), \quad \alpha(t) - 1 \quad (t > \rho_0'') \tag{5.254}$$

are nondecreasing and can take any integral value at most finitely many times. Then it follows from the second equation (5.250) and the two first equations (5.251) that

$$E_1' \subset E_0', \quad E_2' \subset E_0'. \tag{5.255}$$

Comparing (5.252) and (5.253) with (5.255), we obtain

$$E' \subset E_0' \cup W'. \tag{5.256}$$

Since by (3.86) and the second equation (5.250)

$$\text{mes } E_0' = 0, \quad E_0' \subset [-\pi, \pi], \tag{5.257}$$

we have, from (5.94), (5.103) and (5.256),

$$\text{mes } E' = 0, \quad E' \subset [-\pi, \pi], \quad \text{mes } e_0 = 2\pi, \quad e_0 \subset [-\pi, \pi]. \tag{5.258}$$

We now show that

$$\lim_{t\to\infty} \beta_{\alpha(t)}(x) = 0 \quad (x \in [-\pi, \pi] \setminus E', x \in e_0). \tag{5.259}$$

Let x be any point satisfying

$$x \in [-\pi, \pi] \setminus E', \quad x \in e_0. \tag{5.260}$$

Then by the first equation (5.250) there is an integer t_0 such that

$$x \in [-\pi, \pi] \setminus E_t' \quad (t > t_0), \quad t_0 > \rho_2'', \quad x \in e_0, \tag{5.261}$$

whence by (5.249) we have, for the x under consideration,

$$|\beta_{\alpha(t)}(x)| < \omega(t) + \zeta_\tau(x) \quad (t > t_0), \tag{5.262}$$

and then, by (5.92), (5.240) and (5.246), we obtain (5.259).

Now put

$$P_m(x) = Q_t^{(0)}(x) \qquad (M_{t-1} < m \leqslant M_t, t > \rho_2''), \tag{5.263}$$

where $Q_t^{(0)}(x)$ is defined by (5.38). Then since $\rho_2'' \geqslant 3$, by (5.34), the $P_m(x)$ are defined for $m > M_{\rho_2''-1}$, and

$$P_{M_t}(x) = Q_t^{(0)}(x) \qquad (t > \rho_2''), \tag{5.264}$$

and furthermore the sequence of functions

$$P_m(x) \qquad (m > M_{\rho_2''-1}) \tag{5.265}$$

is obtained from the sequence

$$Q_t^{(0)}(x) \qquad (t > \rho_2'') \tag{5.266}$$

by repeating each $Q_t^{(0)}(x)$ a finite number of times.

Since the functions $\tilde{Q}_t(x)$, $t \geqslant 2$, are continuous and therefore finite for $x \in [-\pi, \pi]$ (see the remark after (3.37)), the functions (5.266) and (5.265) have, by (5.34), (5.198) and (5.38), the same property, and moreover the sequences (5.265) and (5.266) have the same limit points for each $x \in [-\pi, \pi]$. Then

$$\overline{\lim_{m\to\infty}} P_m(x) = \overline{\lim_{t\to\infty}} Q_t^{(0)}(x), \quad \underline{\lim}_{m\to\infty} P_m(x) = \underline{\lim}_{t\to\infty} Q_t^{(0)}(x) \quad (x \in [-\pi, \pi]), \tag{5.267}$$

whence, by (5.168),

$$\overline{\lim_{m\to\infty}} P_m(x) = F(x), \quad \underline{\lim}_{m\to\infty} P_m(x) = G(x) \tag{5.268}$$

almost everywhere on $[-\pi, \pi]$.

As before, we shall consider τ to be the function of t defined by (5.91). Then it follows from (3.86), (5.37), (5.93), (5.100), (5.244) and (5.248) that

$$\lim_{t\to\infty} \text{mes}\, E_t' = 0, \quad E_t' \subset [-\pi, \pi] \quad (t > \rho_0''). \tag{5.269}$$

Now we shall consider $t = t(m)$, in turn, to be the function of m defined by

$$M_{t-1} < m \leqslant M_t \quad (t > \rho_0''). \tag{5.270}$$

Then

$$\lim_{m\to\infty} t = \infty, \tag{5.271}$$

and the function $t = t(m)$ is defined for $m > M_{\rho_0''-1}$. Moreover, it follows from (5.269) and (5.271) that

$$\lim_{m\to\infty} \text{mes}\, E_t' = 0, \tag{5.272}$$

and by (3.131), (5.53), (5.54) and (5.271) we obtain

$$\lim_{m\to\infty} \text{mes}\, \tilde{G}_m = 0. \tag{5.273}$$

Comparing (5.259) and (5.271), we also have

$$\lim_{m\to\infty} \beta_{\alpha(t)}(x) = 0 \quad (x \in [-\pi, \pi] \setminus E', x \in e_0). \tag{5.274}$$

In addition, by (5.39), (5.58), (5.59) and the inequality $\rho_0' < \rho_0''$ (see (5.34)), we have

$$|\tilde{B}_m^{(t)}(x, T)| \leqslant K' \cdot |\beta_{\alpha(t)}(x)| + \frac{1}{2^t}$$

$$(x \in [-\pi, \pi] \setminus \tilde{G}_m, M_{t-1} < m \leqslant M_t, t > \rho_0''),$$

whence it follows, by (5.258), (5.271), (5.273), (5.274), and the definition of t as a function of m, that the sequence of functions

$$\tilde{B}_m^{(t)}(x, T) \quad (m > M_{\rho_0''-1}) \tag{5.275}$$

converges to zero in measure on $[-\pi, \pi]$ as $m \to \infty$.

Since it follows from (5.34), (5.35) and (5.50) that

$$2 \leqslant \rho_0'' \leqslant \alpha(t) < t, \quad B'_{\alpha(t),m}(x, T) = B'_{t,m}(x, T)$$
$$(\rho_{\tau-1}'' < t \leqslant \rho_\tau'; \tau = 2, 3, \ldots, \quad m = 0, 1, 2, \ldots), \tag{5.276}$$

we obtain, if we take account of (5.194) and the inequality

$$3 \leqslant \rho_2' < \rho_2'' < \rho_\tau' < \rho_\tau'' \quad (\tau = 3, 4, \ldots) \tag{5.277}$$

(see (5.34)), that

$$T_m^{(1)}(x) = B'_{\alpha(t),m}(x, T) \quad (\rho_{\tau-1}'' < t \leqslant \rho_\tau', M_{t-1} < m \leqslant M_t, \tau = 3, 4, \ldots). \tag{5.278}$$

In addition, by (5.35),

$$t - 1 \geqslant \alpha(t) \quad (\rho_{\tau-1}'' < t \leqslant \rho_\tau', \tau = 1, 2, \ldots),$$

and consequently, by (3.121),

$$M_{t-1} \geqslant M_{\alpha(t)} \quad (\rho_{\tau-1}'' < t \leqslant \rho_\tau', \tau = 1, 2, \ldots). \tag{5.279}$$

By the remark before (5.6), $B'_{t,m}(x, T)$ are special instances of the functions $\tilde{B}_{t,m}(x, T)$, obtained from the latter by appropriate choice of $\tilde{a}_j$ and $\tilde{b}_j$. Then, by (3.157),

$$|B'_{t,m'}(x, T) - B'_{t,m''}(x, T)| < \frac{1}{2^t} \quad (m', m'' \geqslant M_t, t \geqslant 2),$$

and the functions on the left involve a_j' and b_j' only for j such that $1 \leqslant j \leqslant M_t$. If we now replace t by $\alpha(t)$ in the preceding inequality and take account of (5.198), (5.277) and (5.279), we obtain

$$|B'_{\alpha(t),M_{\alpha(t)}}(x, T) - B'_{\alpha(t),m}(x, T)| < \frac{1}{2^{\alpha(t)}}$$
$$(\rho_{\tau-1}'' < t \leqslant \rho_\tau', M_{t-1} < m \leqslant M_t, \tau = 3, 4, \ldots, -\pi \leqslant x \leqslant \pi). \text{ ([12])} \tag{5.280}$$

If we compare the preceding inequality with (5.278), we find that

$$|B'_{\alpha(t),M_{\alpha(t)}}(x, T) - T_m^{(1)}(x)| < \frac{1}{2^{\alpha(t)}}$$
$$(\rho_{\tau-1}'' < t \leqslant \rho_\tau', M_{t-1} < m \leqslant M_t, \tau = 3, 4, \ldots, -\pi \leqslant x \leqslant \pi). \tag{5.281}$$

Then, by (5.38), (5.214), and (5.277), it follows that

$$|Q_t^{(0)}(x) - T_m^{(1)}(x)|$$
$$\leqslant |\tilde{Q}_{\alpha(t)} - B'_{\alpha(t),M_{\alpha(t)}}(x, T)| + |B'_{\alpha(t),M_{\alpha(t)}}(x, T) - T_m^{(1)}(x)| < \frac{4}{2^{\alpha(t)}}$$
$$(x \in [-\pi, \pi] \setminus E_{\alpha(t)}, \rho_{\tau-1}'' < t \leqslant \rho_\tau', M_{t-1} < m \leqslant M_t, \tau = 3, 4, \ldots). \tag{5.282}$$

([12])The left-hand side of (5.280) involves a_j and b_j only for j satisfying $1 < j < M_{t-1}$.

In addition, by (5.29), with r taken to be r_τ, (5.33), (5.35), (5.38) and (5.277), we obtain

$$\beta_{\alpha(t)} = Q_t^{(0)}(x) - B'_{t-1,M_{t-1}}(x, T) \quad (\rho'_\tau < t \leqslant \rho''_\tau, \tau = 3, 4, \dots). \tag{5.283}$$

Moreover, by (5.211) we have

$$|B'_{t-1,M_{t-1}}(x, T) - B'_{t-1,m}(x, T)| < \frac{1}{2^{t-1}} \quad (m > M_{t-1}, t = 3, 4, \dots).\text{ [13]} \tag{5.284}$$

Supposing, as before, that t is the function of m defined by (5.270) for $m > M_{\rho''_1 - 1}$, we take

$$\delta_m(x) = Q_t^{(0)}(x) - T_m^{(1)}(x) \quad (M_{t-1} < m \leqslant M_t, \rho''_{\tau-1} < t \leqslant \rho'_\tau, \tau = 2, 3, \dots), \tag{5.285}$$

$$\delta_m(x) = Q_t^{(0)}(x) - B'_{t-1,m}(x, T) \quad (M_{t-1} < m \leqslant M_t, \rho'_\tau < t \leqslant \rho''_\tau, \tau = 2, 3, \dots). \tag{5.286}$$

Then, by (3.121), which holds for $t \geqslant 2$ (see the remark at the end of §3), the functions

$$\delta_m(x) \tag{5.287}$$

are defined for $m > M_{\rho''_1 - 1}$, since $\rho''_1 - 1 > 1$ by (5.34). We shall show that

$$\lim_{m\to\infty} \delta_m(x) = 0 \tag{5.288}$$

almost everywhere on $[-\pi, \pi]$.

By (5.282) and (5.285), we have

$$|\delta_m(x)| < \frac{4}{2^{\alpha(t)}}$$

$$(x \in [-\pi, \pi] \setminus E_{\alpha(t)}, \rho''_{\tau-1} < t \leqslant \rho'_\tau, M_{t-1} < m \leqslant M_t, \tau = 3, 4, \dots), \tag{5.289}$$

and from (5.277), (5.283), (5.284) and (5.286) we obtain

$$|\delta_m(x)| \leqslant |Q_t^{(0)}(x) - B'_{t-1,M_{t-1}}(x, T)| + |B'_{t-1,M_{t-1}}(x, T) - B'_{t-1,m}(x)|$$
$$< |\beta_{\alpha(t)}(x)| + \frac{1}{2^{t-1}} \quad (\rho'_\tau < t \leqslant \rho''_\tau, M_{t-1} < m \leqslant M_t, \tau = 3, 4, \dots). \tag{5.290}$$

It follows from (5.289) and (5.290) that

$$|\delta_m(x)| < \frac{4}{2^{\alpha(t)}} + |\beta_{\alpha(t)}(x)| + \frac{2}{2^t} \quad (x \in [-\pi, \pi] \setminus E_{\alpha(t)}, t > \rho''_2), \tag{5.291}$$

where, as before, t is the function of m defined by (5.270), and consequently (5.271) is satisfied.

Comparing (5.37) with (5.271), we have

$$\lim_{m\to\infty} \alpha(t) = \infty. \tag{5.292}$$

---

[13] Recall that whenever we do not say for which x a relation is satisfied, it is assumed to hold for all $x \in [-\pi, \pi]$.

In addition, taking account of (5.258), (5.259) and (5.271), we see that

$$\lim_{m\to\infty} \beta_{\alpha(t)}(x) = 0 \tag{5.293}$$

almost everywhere on $[-\pi, \pi]$.

It follows from (5.252) and (5.258) that mes $E_2' = 0$, and then, by the second equation (5.251), (5.271), and (5.291)–(5.293), we conclude that (5.288) is satisfied almost everywhere on $[-\pi, \pi]$.

From (5.50) and (5.285) we have

$$T_m^{(1)}(x) = Q_t^{(0)}(x) - \delta_m(x) + \tilde{B}_m^{(t)}(x, T)$$
$$(M_{t-1} < m \leqslant M_t,\ \rho''_{\tau-1} < t \leqslant \rho'_\tau,\ \tau = 2, 3, \dots), \tag{5.294}$$

and from (5.34), (5.182), (5.194) and (5.286) we obtain

$$T_m^{(1)}(x) = B'_{t-1,m}(x, T) + \tilde{B}_m^{(t)}(x, T) = Q_t^{(0)}(x) - \delta_m(x) + \tilde{B}_m^{(t)}(x, T)$$
$$(M_{t-1} < m \leqslant M_t,\ \rho'_\tau < t \leqslant \rho''_\tau,\ \tau = 2, 3, \dots), \tag{5.295}$$

whence it follows by (5.263) and (5.294) that

$$T_m^{(1)}(x) = P_m(x) - \delta_m(x) + \tilde{B}_m^{(t)}(x, T) \quad (m > M_{\rho_2''-1}). \tag{5.296}$$

Since the sequence (5.275) converges in measure to zero on $[-\pi, \pi]$ as $m \to \infty$, by (5.288) the sequence

$$-\delta_m(x) + \tilde{B}_m^{(t)}(x, T) \quad (m > M_{\rho_0''-1}) \tag{5.297}$$

has the same property.

In addition, by (5.38) and the remark after (3.81), the functions $Q_t^{(0)}(x)$ $(t > \rho_0')$ are continuous on $[-\pi, \pi]$. Then it follows from the definition of the functions (5.265) that these functions are also continuous on $[-\pi, \pi]$. Moreover, since our method T is row-finite, it follows, by (1.20), (1.21), (5.7)–(5.9) and (5.41)–(5.43), that the functions $T_m^{(1)}(x)$, $B'_{t,m}(x, T)$ and $\tilde{B}_m^{(t)}(x, T)$ $(m = 0, 1, 2, \dots;\ t = 2, 3, \dots)$ are trigonometric polynomials, and then we see from (5.285) and (5.286) that all the functions appearing in (5.296) are continuous and therefore finite on $[-\pi, \pi]$. It follows from this equation, from the properties that have been established for the sequence (5.297), and from Lemma 19 of [1], that([14])

$$\overline{\lim_{m\to\infty}}\,(\text{mes}, [-\pi, \pi])T_m^{(1)}(x) = \overline{\lim_{m\to\infty}}\,(\text{mes}, [-\pi, \pi])P_m(x),$$
$$\underline{\lim_{m\to\infty}}\,(\text{mes}, [-\pi, \pi])T_m^{(1)}(x) = \underline{\lim_{m\to\infty}}\,(\text{mes}, [-\pi, \pi])P_m(x). \tag{5.298}$$

We now show that

$$\overline{\lim_{m\to\infty}}\,(\text{mes}, [-\pi, \pi])P_m(x) = F(x), \quad \underline{\lim_{m\to\infty}}\,(\text{mes}, [-\pi, \pi])P_m(x) = G(x). \tag{5.299}$$

([14])Lemma 19 of [1] states the following.

Let the functions $f_m(x)$ and $g_m(x)$ $(m = 1, 2, \dots)$ be measurable and defined almost everywhere on an interval $[a, b]$, and let $\alpha_m(x)$ $(m = 1, 2, \dots)$ be measurable and finite almost everywhere on $[a, b]$; furthermore let the sequence of the latter functions converge in measure to zero on $[a, b]$. Then, if $f_m(x) = g_m(x) + \alpha_m(x)$ $(m = 1, 2, \dots)$, we have

$$\overline{\lim}\,(\text{mes}, [a, b])f_m(x) = \overline{\lim}\,(\text{mes}, [a, b])g_m(x),$$
$$\underline{\lim}\,(\text{mes}, [a, b])f_m(x) = \underline{\lim}\,(\text{mes}, [a, b])g_m(x).$$

It follows from (5.243), (5.122), (5.136) and (5.141) that there is an increasing sequence of integers

$$u_\nu \quad (\nu = 1, 2, \dots), \quad v_\nu \quad (\nu = 1, 2, \dots), \tag{5.300}$$

satisfying

$$\lim_{\nu\to\infty} Q_{u_\nu}^{(0)}(x) = F(x), \quad \lim_{\nu\to\infty} Q_{v_\nu}^{(0)}(x) = G(x) \tag{5.301}$$

almost everywhere on $[-\pi, \pi]$, and

$$u_\nu > \rho_2'', \quad v_\nu > \rho_2'' \quad (\nu = 1, 2, \dots). \tag{5.302}$$

Since (3.121) is satisfied for $t = 2, 3, \dots$ (see the remark at the end of §3), if we put

$$p_\nu = M_{u_\nu}, \quad q_\nu = M_{v_\nu} \quad (\nu = 1, 2, \dots) \tag{5.303}$$

and take account of (5.263) we see that

$$\begin{gathered} p_{\nu+1} > p_\nu > M_{\rho_2''}, \qquad q_{\nu+1} > q_\nu > M_{\rho_2''} \quad (\nu = 1, 2, \dots), \\ P_{p_\nu}(x) = Q_{u_\nu}^{(0)}(x), \quad P_{q_\nu}(x) = Q_{v_\nu}^{(0)}(x) \quad (\nu = 1, 2, \dots), \end{gathered} \tag{5.304}$$

and then, by (5.301),

$$\lim_{\nu\to\infty} P_{p_\nu}(x) = F(x), \quad \lim_{\nu\to\infty} P_{q_\nu}(x) = G(x) \tag{5.305}$$

almost everywhere on $[-\pi, \pi]$.

We have seen that the functions (5.265) are finite for each $x \in [-\pi, \pi]$ (see the remark after (5.266)), and hence, by Theorems A, C, F and G of [1] (§§2 and 3) and (5.268), we obtain

$$\begin{aligned} G(x) &= \varliminf_{m\to\infty} P_m(x) \leqslant \varliminf_{m\to\infty} \left(\text{mes}, \left[-\pi, \pi\right]\right) P_m(x) \\ &\leqslant \varlimsup_{m\to\infty} \left(\text{mes}, \left[-\pi, \pi\right]\right) P_m(x) \leqslant \varlimsup_{m\to\infty} P_m(x) = F(x) \end{aligned} \tag{5.306}$$

almost everywhere on $[-\pi, \pi]$.

Since, by (5.304), the sequences

$$p_\nu \quad (\nu = 1, 2, \dots), \quad q_\nu \quad (\nu = 1, 2, \dots) \tag{5.307}$$

defined by (5.303) are increasing, if we again apply Theorems A, C, F and G of [1], Lemma 5 of [7] (§2) and (5.306), we conclude that

$$\begin{aligned} \varliminf_{\nu\to\infty} P_{p_\nu}(x) &\leqslant \varliminf_{\nu\to\infty} \left(\text{mes}, \left[-\pi, \pi\right]\right) P_{p_\nu}(x) \leqslant \varlimsup_{\nu\to\infty} \left(\text{mes}, \left[-\pi, \pi\right]\right) P_{p_\nu}(x) \\ &\leqslant \varlimsup_{m\to\infty} \left(\text{mes}, \left[-\pi, \pi\right]\right) P_m(x) \leqslant F(x), \\ G(x) &\leqslant \varliminf_{m\to\infty} \left(\text{mes}, \left[-\pi, \pi\right]\right) P_m(x) \leqslant \varliminf_{\nu\to\infty} \left(\text{mes}, \left[-\pi, \pi\right]\right) P_{q_\nu}(x) \\ &\leqslant \varlimsup_{\nu\to\infty} \left(\text{mes}, \left[-\pi, \pi\right]\right) P_{q_\nu}(x) \leqslant \varlimsup_{\nu\to\infty} P_{q_\nu}(x) \end{aligned}$$

almost everywhere on $[-\pi, \pi]$, whence by (5.305) we obtain (5.299). Comparing the latter equation with (5.298), we obtain (1.19).

Thus we have defined a trigonometric series (1.5) satisfying (1.14) of Theorem 1 (see the beginning of §4).

Moreover, at the beginning of §4 we chose arbitrary measurable functions $F(x)$ and $G(x)$ satisfying (1.13) almost everywhere on $[-\pi, \pi]$, and we have defined a subseries (1.18) of (1.5) for which (1.19) and (1.22) are satisfied (see the end of §4). Consequently the series (1.18) satisfies condition a° (see the statement of Theorem 1 in §1), and the proof of this theorem is complete (see the remark at the beginning of §5).

Bibliography

1. D. E. Men'šov, *On convergence in measure of trigonometric series*, Trudy Mat. Inst. Steklov. **32** (1950); English transl. in Amer. Math. Soc. Transl. (1) **3** (1962).

2. ______, *On limits of indeterminacy in measure and limit functions of trigonometric and orthogonal series*, Dokl. Akad. Nauk SSSR **160** (1965), 1254–1256; English transl. in Soviet Math. Dokl. **6** (1965).

3. ______, *The limits of indeterminacy in measure of the T-means of subseries of a trigonometric series*, Some Problems of Mathematics and Mechanics (M. A. Lavrent'ev Seventieth Birthday Volume), "Nauka", Leningrad, 1970, pp. 198–204; English transl. in Amer. Math. Soc. Transl. (2) **104** (1976).

4. ______, *Limits of indeterminacy in measure of T-means of trigonometric series*, Mat. Sb. **81 (123)** (1970), 485–524; English transl. in Math. USSR Sb. **10** (1970).

5. N. O. Sinanjan, *Limits of indeterminacy in measure for the T-means of series in complete orthogonal systems*, Izv. Akad. Nauk Armjan. SSR Ser. Mat. **5** (1970), 522–533. (Russian)

6. D. E. Men'šov, *On limit functions of a trigonometric series*, Trudy Moskov. Mat. Obšč. **7** (1958), 291–334; English transl. in Amer. Math. Soc. Transl. (2) **111** (1978).

7. ______, *Limits of indeterminacy in measure of trigonometric and orthogonal series*, Trudy Mat. Inst. Steklov. **99** (1967); English transl., Proc. Steklov Inst. Math. **99** (1967).

ABCDEFGHIJ–AMS–8987654321